D1105523

Panofsky on Physics, Politics, and Peace

Pief Remembers

Wolfgang K.H. Panofsky

Panofsky on Physics, Politics, and Peace
Pief Remembers

Contributing Editor: Jean Marie Deken

 Springer

Wolfgang K.H. Panofsky
Stanford Linear Accelerator Center
2575 Sand Hill Road
Menlo Park 94025
pief@slac.stanford.edu

Jean Marie Deken
Stanford Linear Accelerator Center
2575 Sand Hill Road
Menlo Park 94025
jmdeken@slac.stanford.edu

Cover illustration: The author during the first trip by U.S. high energy physicists to the Soviet Union, 1956. (Credit: Panofsky Family Collection.)

Library of Congress Control Number: 2007927311

ISBN: 978-0-387-69731-4   e-ISBN: 978-0-387-69732-1

Printed on acid-free paper.

9 8 7 6 5 4 3 2 1

springer.com

Preface

This volume contains an "unsystematic account" of my past work; it is not intended to be an autobiography in the conventional meaning of the term. It is not even remotely a scholarly description of the momentous developments in which I was able to participate; rather it is a recital of "memorable" episodes, borrowing from the "compulsory preface" of a facetious British history:[1] "History is not what you thought. It is what you can remember." Thus this volume suffers from many "sins of omission," including full attribution of deserved credits, but, it is hoped, only few "sins of commission." The author is greatly indebted to his colleagues and his wife, Adele, who kindly reviewed many segments of the manuscript describing shared experiences. They are Sidney Drell, Greg Loew, Ed Lofgren, Harvey Lynch, Richard B. Neal, Richard Panofsky, and Burt Richter. But the author, needless to say, is responsible for any errors.

Because of the multitude of topics into which I was drawn concurrently, a strictly chronological account would prove unreadable. Accordingly the book is divided into chapters, each of which covers a limited period of engagement in a coherent subject matter; an approach clearly again unsystematic but hopefully more conducive to conveying the substance of the work.

This account does not include a description of my family life. Although I was voted "Most Likely Bachelor" by my college class, I have been married since 1942 to a unique and wonderful lady. We contributed to the population explosion and I have presided over ten marriages of my five children. Description of the many family activities and our joint learning is not included in this volume.

The manuscript was prepared by Ms. Ellie Lwin to whom I owe great gratitude. She has the great talent of recording what I meant to say, not what I actually literally presented. Moreover, she dedicated an heroic effort toward preparing the index. Gregory Stewart has converted the pictorial material into forms suitable for use as figures in the text. And then the Contributing Editor, Ms. Jean Marie Deken, edited the entire volume and, with her associates in the SLAC Archives, succeeded in nailing down facts and dates when my memory or records were insufficient.

Table of Contents

1
School Time in Germany

It is customary for most biographies of a scientist to open with an account of how his scientific interest was stimulated by creative teaching of science in school or by encouragement of his scientific interest by his family. I'm sorry to report that neither was true in my case. Neither my elementary school nor high school (a German Gymnasium) offered any significant courses in science during my attendance, and my parents were scholars in the history of art who described their two children as *Klempners* (plumbers) when their interests in scientific subjects became manifest.

I was born in Berlin on April 24, 1919, the second offspring of my parents Erwin Panofsky and Dorothea Panofsky (born Mosse). My parents met at the art history seminar conducted in Berlin by Professor Adolph Goldschmidt, a famous art historian of his generation. My father was the son of a successful businessman. My mother was the daughter of a Prussian jurist who was one of fourteen siblings and who had reached the highest level of judicial seniority possible for a Jew at the time. Both my mother's and father's families included some very prominent individuals. For instance, my mother's father, Albert Mosse, was assigned to help the Japanese government to generate a constitution and to establish laws governing cities and prefectures during the Meiji Restoration. One of his many brothers, Rudolf, was the publisher of the *Berliner Tageblatt*, the leading newspaper in Berlin. My aunt, Martha Mosse, was the first woman to serve as police commissioner of the City of Berlin. My father was possibly the most eminent art historian of his time. Even today, articles commemorating his contributions appear in the German press. His letters (27,000 of which are preserved) have been incorporated in part in a five-volume publication of Professor Dieter Wüttke,[1] of which three have appeared to date. The letters written from Japan by my maternal grandfather and grandmother have been published under the title (translated from the German) *Almost as My Own Fatherland.*[2]

My father was educated in Hannover, Germany, but moved to Berlin where he married my mother, a fellow student. He started his postgraduate studies in Freiburg, and wrote a highly prominent thesis on Dürer's art theory. He subsequently received offers from both the then-leading German University at Heidelberg and from the University at Hamburg, and accepted the latter, moving to Hamburg in 1920 before I was one-year old. He continued his scholarly work based in Hamburg from that time until the Nazis forced him from Germany in 1934. At Hamburg he was associated with several institutions: the University, the *Kunsthalle* (Hall of Arts), and the renowned Warburg Library; he enhanced the stature of all these institutions.

In consonance with the above, I grew up in a typical middle-class but scholarly academic household. My parents traveled a great deal, conducted seminars in the evening in our home, and many professional colleagues were constant visitors. The care of their children was partially the job of a live-in maid, Bertha Ziegenhagen. She was a wonderful person and correspondence with her after my family emigrated to the United States is contained in the publication of my father's letters.[1]

Although my parents had no expressed interest in science, they were fully aware of the growing preoccupation of their children with technical matters. My father liked to pretend his total lack of interest in family affairs; he once said facetiously he wished to have nothing to do with his two children until they could speak fluent Latin. He taught me chess when I was four but ceased playing with his children when they started winning.

After completing grade school, I founded a *Bastelverein*, or tinkerers club, with some fellow students. We gathered materials to make all sorts of things. Our principal toy was a Märklin construction set, which my parents supplemented annually, greatly increasing the total inventory of parts. It was actually remarkable that we did not kill ourselves, because many of our "toys" were powered directly from the wall plug using a homemade exposed voltage divider. Among the many things we made was an automatic vending machine to sell candy and cigarettes at exorbitant prices to my father's art history students at their seminars. Our family had an extensive library, and there were some technical books. Of interest to me was a book entitled *Physical Playbook*[3] which showed us the way to have fun with technical things. Also, the library contained a book on the theory of sound which was given to my father as a prize when he was a young student, and there were encyclopedias where one could find technical subjects.

I went to a German private elementary school for four years, and then my father enrolled me in the *Johanneum*, a Gymnasium that had been founded over 400 years ago, at the time of the Protestant Reformation. When I was attending school, as is still mostly the case today, all German educational institutions from the fifth grade onward were segregated into different educational tracks. Thus, parents had to choose, while their children were quite young, whether the children should be given a highly classical or a more applied education. The *Johanneum* was very classical indeed; its teachers were generally associated with the University and continued to publish scholarly articles. Latin was taught throughout, but no science or modern foreign language courses were given, except during the last one or two years before graduation, and I departed before reaching that stage.

I participated in no religious activities while in Germany, and such participation on the part of my family was infrequent. At the Gymnasium, religious history was taught as part of the general history sequence. That instruction did an excellent job of delineating the influence of religion on the evolution of civilization in the past, but such instruction was entirely secular in character. I still retain an atlas given to me at that time describing the changing boundaries as influenced, to a large extent, by religion.

Music was taught extensively at the Gymnasium. All students sang in the choir, and the school auditorium incorporated a large pipe organ. I remember the 400-year anniversary of the Gymnasium. We performed in many musical presentations. On one occasion, we simply failed to start singing when directed, to the great fury of our music teacher. On the occasion of the anniversary, the teachers published a commemorative volume of scholarly contributions. One of those articles traced the origin of the family names of all pupils, including the one of my brother and myself. However, in our case, the authors got it wrong.

Our life was without major hardship, but continued to be touched by the political upheavals in Germany. In 1923, inflation made money dive in value precipitously as soon as it was obtained, so my mother had to rush to the countryside to buy eggs, vegetables, and meat just as soon as my father received his pay. Postage stamps reached denominations in the billions of marks. But then came the Nazis. Hamburg was one of the last German cities to yield to them. I remember pamphlets being distributed entitled *Haltet die Tore offen* ("Keep the Gates Open"), emphasizing the existing commercial and cultural relationships to other countries, particularly England. However, school became increasingly militarized; our school excursions changed from nature walks to enforced marches carrying heavy backpacks. Students had to reciprocate the teachers' greetings of "Heil, Hitler!" Sometimes we did this with so much false enthusiasm that the poor teachers were unable to lower their arms, even for a short period of time.

Jews were banned from physical education. Apparently mental contact with Aryans continued to be accepted, but joint physical exercises were perceived to be contaminating. Thus, the non-Aryan pupils joined a separate private Jewish athletic society where we played soccer and other games. In spite of this situation, I participated in athletic competitions sponsored by the city of Hamburg, including the *Alster-Staffel*, a relay race circumnavigating Alster Lake around which Hamburg is built.

My parents' effort to interest my brother and me in art was a total failure. We were taken to museum tours where my parents spent as long as four hours in front of a single picture. There is probably no surer way to divert your children's interest from art to "plumbing."

I had much contact with classical music, which flourished in Hamburg. My brother and I went to many concerts, and I had piano instruction from a lady living far from the center of town, where we lived. Beyond these lessons, my own musical activities were minor: they included conducting the

Haydn Toy Symphony, and playing percussion in the school orchestra. But our family's interest in music remained intense. My father was extensively involved in arranging chamber music concerts in Hamburg; and when we had to leave Germany in 1934, the wind section of the Hamburg Philharmonic Orchestra performed a wonderful Mozart goodbye performance in our house; a very sentimental occasion.

My father accepted a one-half year position as a visiting professor at New York University in 1932 and 1933, and he wrote many letters to his wife and sons in Hamburg, describing his experiences (including Prohibition!) in the United States. He thus had "one foot across the ocean" even before he was formally dismissed from his professorial position in 1934. His correspondence[1] documents in detail the totality of that transition. The faculty at the University of Hamburg was compelled—against their judgment—to vote for his dismissal, but concurrently expressed regrets in private. At the same time, the major institutions with which my father was associated faced inevitable expulsion or neglect from the Reich. After the agonizing search for a new location, the Warburg Library moved all their famed collections to England, and the Art Historical Seminar in Hamburg,[4] which my father conducted, essentially collapsed.

Throughout my time in Hamburg, my family's monetary resources diminished. My father was not paid a real salary until he became a professor in Hamburg, and even thereafter, that salary did not cover expenses. I remember the frequent occasions when it was decided to liquidate what savings there were. This turned out to be a good thing: when our family left Germany; the Nazis did not permit emigrants to take any money with them. But this was no problem because essentially none was left.

My father had very little contact with the science professors at the university. The exception was Professor Otto Stern, who received a Nobel Prize in Physics. He was also forced to leave Germany, and we met him again much later in Berkeley, California. Stern was never able to resume productive work, but this was not the case with my father, who adapted extremely well to the American way of life, as described in the next chapter.

2
Transition to the United States and Undergraduate Life at Princeton University

The dismissal of my father made continued stay in Germany impossible, although at the time of our emigration in 1934, actual persecution of Jews was not yet extensive. The Warburg Library, now located in London, offered a position to my father. At the same time, he was also offered a teaching appointment—at substantially higher compensation—at New York University, where he was already well known as the result of his earlier visits. But the main problem of how to provide for his sons' educations remained. Charles Rufus Morey, the chair of the Princeton University Art Department, proposed a solution. He offered my father free housing for the family along Princeton's Prospect Avenue and free tuition for his two children, in exchange for some teaching concentrated in two days of the week. My father accepted this offer, which resulted in a complex arrangement that left him time to commute to New York University, to which he would dedicate the largest share of his teaching attention.

While my parents prepared for the big move to the United States, my brother and I were deposited in a small fishing village in England. The stay in England gave us an opportunity to learn a bit of English. After reuniting briefly in England, the family crossed the Atlantic Ocean by ship, and was received in New York Harbor by friends from Princeton, and introduced to our new home. My parents filed "first papers" to obtain U.S. citizenship shortly after arrival and were assured, incorrectly, that they and I would become U.S. citizens before I reached my 21st birthday. This failure to apply separately for my citizenship at that time would complicate my life a bit later.

We decided that, despite our young age (I was 15 and my brother was 16) we would enroll at Princeton University without attaining a formal high school diploma in the United States. We concluded that the social tensions inherent in joining a high school class would be more serious than the ones we would encounter as entering students at the university. I remember well the interview with Dean Radcliffe Heermance, the dean of admission to the university. We recited our curriculum at the German Gymnasium, and after some deliberation, clearly influenced by Professor Morey's prior commitments to my father, he admitted both of us "on probation" to the freshman class.

I enrolled largely in technical courses in addition to a further course in Latin. I avoided courses in English or history because, with my poor English, I did not think I could cope with the heavy load of essay writing. I recall meeting Dean Heermance during a walk on campus, and he asked me how I was doing. This was right after concluding the first term of the freshman year. I said I got one "2", equivalent to a "B"; I never got a "2" again. The Dean was a bit perplexed about this result of probation.

Life at Princeton proved intense and busy. My brother and I slept at home, but otherwise participated fully in undergraduate life. We had few friends. Most of the other undergraduates considered my brother and myself as strange phenomena, but exhibited tolerance. My full name proved unpronounceable to most, so one of my nicknames Piefke, or Pief for short, stuck to me throughout life. Piefke and Paffke (used for my brother Hans) derive from German cartoons; in general, Piefke was a slightly derogatory term applied to bourgeois Germans, particularly those traveling abroad. Piefke was also a Prussian composer of military music, as well as a figure in one of Wilhelm Busch's famous satirical poems and sketches.[1]

One of my friends was Raymond Emrich, one of the few of my classmates interested in physics. We became close friends and kept in touch throughout our careers. He recently died while emeritus professor at Lehigh University. I used to go to the movies with him and another friend, both of whom towered over my five-foot, two and one-half inch height. On those occasions, Raymond would go to the cashier and ask for a "one and a half" ticket.

Grades were quite important to most students at that time. They were publicly posted at the end of each term, with students gathering to inspect how they did. There was not too much social life on campus. A fair amount of student activities concentrated on transferring money to the less affluent students on campus from those who were more fortunate. I had a job as usher at the football games, and later supplemented my funds extensively through tutoring. The only thing I remember from ushering was that I once received a comment, "You are the worst usher I've ever met." In contrast, tutoring became quite successful: I even recall that one of my physics professors called me in to inquire whether I had access to the examinations to be given (I didn't) because one of my disciples suddenly did unexpectedly well.

During my Princeton stay, I was exposed to some inspiring and able instructors. My freshman physics course was given by Henry DeWolf Smyth. He later became very well known as the author of the Smyth Report describing the work of the Manhattan District during the war, and was also the first American delegate to the International Atomic Energy Agency. I took quantum mechanics from John Wheeler, and there was a comprehensive physics curriculum with small classes.

Both my brother and I learned how to drive, thanks to the instruction given us by my friends. At the end of our first year at college, we decided to buy a 1926 Buick touring car and use it to see America. We traveled in the Buick with another student friend all the way from Princeton to the West

Coast, camping out or staying in cheap cabins. My father made arrangements for us to stay with one of his former students in California, who in turn showed us around Stanford University and took us to Lake Tahoe. On the way back we stayed for a few days in the Black Hills of North Dakota, joining a geological exploration headed by a professor who was a neighbor at Princeton. I don't believe my parents fully realized the total implication of a trip approaching 10,000 miles in all during one summer for two inexperienced kids with one slightly more experienced American fellow student, but all went fairly well, although the roof of the car blew off in Iowa on the way home.

Between my sophomore and junior years, I was able to get a summer job at the RCA tube division in Harrison, New Jersey, together with my friend Raymond Emrich. We commuted by "Hudson tube" from a New York apartment. This was my first exposure to an American industrial laboratory. I was given an assignment to examine the dependence of thermionic emissions from oxide cathodes on the concentrations of strontium and barium in these cathodes. Most of the work consisted of preparing a variety of such cathodes, having those incorporated in production line tubes, then measuring their emission, and smashing the tubes to bits. The work resulted in my first publication.[2] It was a valuable experience. I was surprised that these tubes were all in full-scale production without such measurements having been made; I also was surprised to see in the production line that RCA was manufacturing tubes with the labels of other, supposedly competitive, American companies.

At the time of my attendance, Princeton offered real research opportunities to its undergraduates. I produced a junior thesis[3] on the vibrations of a piano string, which involved an extensive literature search and a complete Fourier analysis of a struck vibrating string. In my final year, I wrote a senior thesis under the direction of Professor Walker Bleakney on radiation measurements using a high-pressure ionization chamber.[4] The radioactive isotopes measured were produced in the small cyclotron nearby in the basement of the Palmer Laboratory. It was built under Professor Milton White, with whom I interacted much later in connection with what is now the Fermi National Accelerator Laboratory. This work was very educational, because I was able to spend many hours in a small basement room building the radiation detection apparatus from scratch. Inasmuch as few supplies were directly available, I had to learn a great deal of machine work and I even wound my own transformers for the power supply. The experience also gave me the opportunity to become acquainted with the workings of the cyclotron. In addition, during this period I became an assistant to Professor Henry Eyring, a physical chemist and at the same time a Mormon Bishop. Eyring wrote papers on the theory of liquids, and I wrote lecture notes for his course on crystallography. I remember many sessions with him where he would alternately lecture me on the theory of liquids and the evils of drink, all the while chomping on a big cigar.

In addition to studies in physics and chemistry, I took French and some Latin literature courses. Interestingly enough, my brother, who was more competent in Latin and Greek than I, won a very large, prestigious endowed scholarship after a competitive examination in Latin. To the great chagrin of the classics department, he promptly became an astronomer. I received the "Wood Legacy Prize" as the "Best Member of the Junior Class."

Princeton University had compulsory physical education. After being exposed to a frustrating "body-building" class, I enrolled in basketball. I joined the lowest ranked team, and as a five-foot-two inch basketball player, annoyed my "giant" opponents by dribbling basketballs through their legs.

My days at Princeton taught me a great deal about American society, although of course we lived in a relatively restricted environment. I was exposed to the racial problems of the time. I once had lunch with a fellow student from the South at a local restaurant, and I asked him what he would do if a Negro came in and sat down at one of the tables. He said, "I would leave." I asked why, and he said, "Superiority." This was the end of our acquaintance.

I also learned about some of the social disparities in American society. Princeton was partially a fairly affluent town composed of the university community and residences of business commuters to New York City, but it also encompassed a largely black ghetto with poor inhabitants. The majority of Princeton students, after surviving an elaborate "bickering" selection process, joined "eating clubs" which required fairly high fees. I organized an enterprise for collecting the very large quantities of food left over from the eating club lunches and delivering it to the largely black communities, who greatly welcomed the distribution.

Princeton University, at the time of my attendance, had "compulsory chapel." All students had to select the religion under which they wished to participate in services. Princeton University had an imposing central church in which Protestant services were offered on Sundays. Very few students initially subscribed to Jewish services, which were conducted by an atheist graduate student who obtained literature to conduct the service from a synagogue in New York. Then to our amazement, the Jewish congregation kept growing and growing, greatly exceeding the small number of Jewish students at Princeton. A rabbi had to be imported from New York to conduct the services. The University administration finally caught on to the situation, realizing that the motivation of the students was to attend the Friday services in order not to spoil the weekend. As a result, a rather absurd rule was instituted that Christians could not attend Jewish services, although Jews were free to do the inverse.

My father terminated his complex dual appointment at Princeton and New York Universities, and late in 1935 became the first member of the School of Humanities at the Institute of Advanced Studies (at Princeton). In 1938 our family moved to a new home designed by a Princeton faculty colleague and built on the lands of the Institute. I observed the planning and construction process of the house with great interest.

At the Institute of Advanced Studies, my father became a colleague of some of the great physicists and mathematicians then working or visiting there. He became an interlocutor with Albert Einstein and Wolfgang Pauli, both of whom showed great interest in comparing phenomena in nuclear physics with mythical, medieval, and Greek concepts. There is interesting correspondence between Pauli and my father comparing the neutrino, then conjectured by Pauli, with the medieval image of angels. I was much too junior to participate in these discussions, so my only role was that of chauffeur for my father and these senior physicists. I recall driving along Route 1 near Princeton, with Einstein and my father talking in the back seat. A traffic cop stopped us. I was afraid that something was wrong with my driving, but the cop said, "I just wanted to look at the great man."

I graduated in 1938 with "highest honors" and was elected as salutatorian of my class. In that role, I had to present a Latin speech at graduation. I well remember writing it, including encountering difficulty in translating "behind the eight-ball," the symbol of our class, into Latin. I was also voted "Most Brilliant" and "Most Likely Bachelor" by my class, a judgment contradicted by events identified in the next chapter.

3
Graduate Study and War Work at Caltech

After graduation from Princeton University in 1938, I submitted several applications for graduate school. I had a very interesting visit at Columbia University, including a stimulating discussion with I. I. Rabi. I received a multipaged single-spaced personal letter from R. A. Millikan, the president of Caltech (formally the chairman of the Executive Council of the institute), extolling in great detail Caltech programs and opportunities, and offering me a teaching assistantship. In response, I decided to transplant to the West Coast; because there was plenty of time to get there, I decided to travel by freighter, via the Baltimore Mail Line through the Panama Canal to Los Angeles. It was a thoroughly interesting trip with several stops along the American and Mexican coasts. After arriving in Pasadena, I received lodging at the Athenaeum, the faculty and graduate student facility at Caltech. I lived first in the Loggia, a single open space under the roof with showers at both ends for the resident graduate students. It has since been condemned as unsafe because of the single stairway leading to the lower floors. Later, as part of a scholarship, I was assigned a single room.

Caltech offered excellent opportunities for the intellectual development of graduate students. It was required to pass a preliminary examination before undertaking research, and therefore I undertook extensive course-work. I shared an office with Donald Wheeler, who later taught at Lehigh University in Bethlehem, Pennsylvania. Course work at Caltech was heavily problem-solving oriented, and was taught by great people who combined excellent teaching with various idiosyncrasies. Let me give some examples.

Ira Bowen taught an excellent course in optics, but unbeknownst to his students, was one of the senior astronomers and instrument designers at the Mount Wilson Observatory in the hills behind Pasadena.

The British mathematician Harry Bateman taught a very challenging course in higher mathematical analysis. I recall that he gave a final examination consisting of about a dozen problems with the admonition, "Do not do more than nine." I solved two, and got an "A" and found later that the questions were based on Bateman's work which he subsequently published in mathematical journals.

Paul Epstein was a classical theoretical physicist who had made seminal contributions to nonlinear hydrodynamics. He taught an excellent course in theoretical electricity and magnetism and another course in thermodynamics. He would lecture pacing rapidly across the full width of the lecture room in a motion that we students called "Epicycles."

In contrast, William Smythe taught a purely problem-oriented electricity and magnetism course that continued the tradition set by the famous Tripos examinations in Great Britain. His course, which is reflected in the textbook he published, was considered one of the major hurdles for a graduate student at Caltech. My fellow graduate student, Charles Townes (later to receive the Nobel Prize as co-discoverer of the laser) and I boasted of having solved all of Smythe's problems.

Fritz Zwicky, well known for his identification of supernovae, taught mechanics. His teaching style was highly temperamental. He called his graduate students to solve problems on the blackboard; if he felt their performance was inadequate, he would follow their presentations with an eraser, rubbing out the formulas as soon as they were written. I recall that in my final Ph.D. examination, he asked me, "What am I thinking about now?" I replied, "I rightly don't know."

Richard Tolman taught very rigorous courses in statistical mechanics and general relativity. I recall that he gave a true–false examination in general relativity, a feat very difficult to accomplish without ambiguities! There was only a brief course in nuclear physics taught by Charles Lauritsen, and another course in atomic physics taught by Robert Millikan.

Quantum mechanics was taught by Linus Pauling of the chemistry department, following the general lines of his well-known book, *The Nature of the Chemical Bond*. It was a very interesting course of applications, but not on the theory of quantum mechanics. My difficulty was that the chemical phenomena which he introduced to be explained by quantum mechanics were largely unknown to me. A brief course in solid-state physics was given by William V. Houston, who soon thereafter departed for Houston, Texas. It is worth noting that there were no advanced courses in quantum mechanics, or in solid-state physics, or any courses giving deeper insight into elementary particle physics. Thus, although we graduate students received excellent analytical and problem-solving instruction and training, we had to learn independently the background of those topics which we chose to pursue in our research work.

Caltech gave us full responsibility for the teaching of undergraduate sections. Because the students in the different sections received identical examinations, the graduate student instructors would contest their student contingents like gamblers at a horse race, betting on whose section would get the best grades. All graduate students ate lunch together at the Athenaeum, and therefore we were able to compare experiences with different instructors doing research and give guidance to fellow students who had not embarked on that phase of their work. The Athenaeum enforced a formal dress code requiring coats and ties, even at lunchtime, as mandated by Mrs. Millikan. Reserve ties were kept in the cloakroom, and enforcement of the code was left to the waitresses. I recall one of my fellow students with a long black beard challenging the waitress to lift his beard to check whether he wore a tie. She refused.

After passing my qualifying examinations, I chose to do research with Professor Jesse W. M. DuMond. He conducted experimental research on x-ray physics two floors below my assigned office, and I had the opportunity to drop in there frequently before deciding to join his research group. DuMond was an extremely universal physicist whose work spanned from classical theoretical analysis through design of instruments to experimentation and hands-on fabrication. He had incomparable ability in geometrical design. I do not elaborate further on his work here; after DuMond died in 1976, I wrote his scientific biography for the memoirs for the National Academy of Sciences.[1] At the time I joined his group, DuMond's research work centered on what was known as the "Watters generator," a very high-powered x-ray tube assembled largely of surplus parts and named after the donor of research funds for DuMond's work. In those days support had to be solicited from private funds or through university endowment, as government funds for such work were nonexistent.

The variety of research undertakings using the Watters generator involved a succession of graduate students, resulting in a number of theses through partially collaborative efforts. My thesis topic was a precision measurement of the ratio of Planck's constant to the charge on the electron. The work largely used existing equipment, but I designed and built a precision voltage regulator, a highly accurate voltage divider, and devices to clean low atomic number deposits from the anode of the x-ray tube. These latter deposits derived from the oil diffusion pumps generating vapors that were not perfectly removed by liquid nitrogen traps. Some of the equipment had been designed by DuMond and his previous collaborators. The wavelength of the x-rays was measured by a two-crystal (calcite) spectrometer of beautiful mechanical design which was built by Douglas Marlowe. The method of measuring h/e consisted of measuring the endpoint of the continuous spectrum of x-rays produced by bombardment of electrons of about 20 KeV, to a precision unprecedented at the time.[2] Figure 3.1 shows the final spectrum. The accuracy of the determination, about 3 parts in 10^4, was a record at the time, but has of course since that time been superseded by several orders of magnitude. Because of the finite resolution of the spectrum measurements and the finite precision of the voltage measurement, I developed a new method of unfolding the resolution curves from the measurement to lead to the final estimated precision.

DuMond incorporated the measurements of h/e into a geometrical representation named the isometric consistency chart. That representation exhibited values of the charge and mass of the electron, and of Planck's constant, along the sides of a hexagon. It plotted each measurement and its assigned

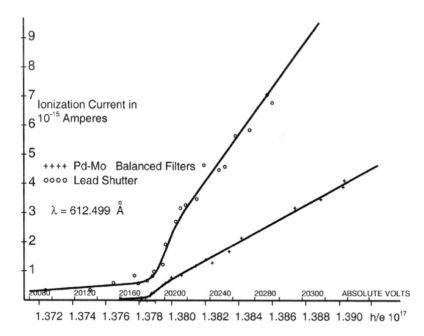

FIGURE 3.1. The continuous x-ray spectrum near threshold. (From J. W. DuMond, W. K. H. Panofsky, and A. E. S. Green.[3])

standard deviation as lines of appropriate slope on the diagram. DuMond and his associate E. Richard Cohen continued to be two of the "guardians" of the natural constants, publishing such analyses for some time.

DuMond, in addition to the high-voltage x-ray tube described previously, designed an ingenious curved crystal spectrometer for high-precision wavelength measurement of high-energy gamma rays, rather than x-rays. The instrument was based on bending a thin quartz crystal plate to an appropriate radius of curvature, and then mounting the source of gamma rays and the detector on mechanisms designed so that they were located at the reciprocal foci produced by Bragg scattering of the gamma rays from the crystal. These motions were controlled by precision-lapped screws. In accordance with the tradition of DuMond's laboratory, I collaborated with subsequent graduate students and other physicists in bringing this instrument to fruition,[4] notwithstanding the interruptions of the war. Interestingly enough, this instrument, although of unprecedented precision, suffered from its small geometrical acceptance. Therefore, its utility depended in practice on the availability of very strong radioactive sources produced in the nuclear reactors that the advent of nuclear energy during the war made possible. This γ-ray spectrometer remained in productive use at Caltech long after I left.

DuMond's family was highly hospitable, and his graduate students were frequent guests at their home. I became well acquainted with the family members, including DuMond's daughters, aged 12 and 16 at the time I started my research. Aside from enjoying Professor DuMond's hospitality, social life consisted largely of excursions with our fellow graduate students. Many expeditions were organized to climb the major peaks in the neighborhood, in particular Mount San Jacinto and San Gorgonio. Friendships developed among many of the graduate students and I kept in touch with many of them who are now well-known physicists.

During the later parts of my thesis work, the clouds of war began to overshadow much of the work at Caltech. DuMond left frequently for the East Coast to engage, among other projects, in problems of undersea warfare. The central military program at Caltech was the development of rockets under the direction of C. C. Lauritsen. Some of these rockets were also used as targets for the training of troops for anti-aircraft fire. DuMond and colleagues observed some of these exercises using rockets as well as sleeves towed by airplanes as targets. He was disappointed that any hits on the targets by the fired bullets were extremely rare, and that therefore the statistical significance of assessing the performance of the anti-aircraft batteries was exceedingly poor. He thus proposed that the target incorporate a device that could sense the trajectory of the passing bullets from a longer distance, thereby effectively constituting a bull's eye in the sky. He consequently formed a research group whose work was sponsored by the U.S. government's National Research Defense Council (NRDC) to undertake this task.

The first attempt to do so employed magnetic detection by magnetizing the bullets and detecting their passage through coils mounted on the targets. The method proved unsatisfactory because of the excessive magnetic noise environment and the rapid fall-off of the magnetic field with distance. We then switched to the detection of the shockwaves produced by the supersonic speed of the bullets. We worked on both the theory of shockwave propagation[5] and the actual design and fabrication of shockwave detectors with a flat frequency response from 0 to well above 10 KHz. These detectors consisted of condenser microphones directly frequency-modulating an oscillator. Two of these microphones were embedded in a spherical housing so that the sum of the signals of the two microphones resulted in a nearly spherical acoustic sensitivity pattern, thus giving a measurement of miss-distance. The difference between the signals would identify the direction of the miss. This device, called the Firing Error Indicator (FEI), proved quite successful at not-too-high target speeds, and was put into commercial manufacture. A large number of field tests had to be performed at various military test ranges, including nearby ranges at Camp Irwin in the Mojave Desert, but also at ranges in Laredo, Texas, and in Virginia in the East.

In addition to the military research and development carried out at Caltech, the Institute provided extensive refresher courses for high-ranking military officers and engineers working in the defense industry. I taught several of these courses, mostly in the evenings. In parallel with these activities, there developed increasing hysteria about a possible Japanese invasion of the

West Coast. California was designated a potential war zone under the command of a General DeWitt. In turn, the Enemy Exclusion Act was passed that provided for the potential relocation of any enemy aliens from California to detention camps in the desert. In practice, this Act, later declared unconstitutional, was racist, being only enforced against Japanese, including U.S. citizens of Japanese descent. However, Italian and German citizens were subjected to several restrictions, including not being permitted to travel more than five miles from their domicile without permission, and being subjected to a curfew requiring that they retire before 9 PM. I recall hiking and canyon climbing with DuMond's elder daughter, Adele, in the hills behind Pasadena armed with a map indicating a circle of five-mile radius around Caltech (which was engaged in secret military work!). Hopefully, we knew where we were going relative to that map. I also recall teaching one of the evening courses when a burly military policeman appeared asking why I wasn't in bed, to the great astonishment of the audience.

Then came Pearl Harbor. I learned about the attack while listening to a car radio with Adele in the hills. I had been teaching military officers, all two or three times my age, in Caltech style, asking them to work problems on the blackboard and generally bossing them around. After Pearl Harbor, they were ordered into uniform and they all turned out to be generals, to my great embarrassment.

The research and development on the shockwave devices and other military programs required security clearances; I received my clearance despite being "an enemy alien." This was not too uncommon an occurrence at the time, as I later found out when starting work in connection with the Manhattan Project.

Despite the diversion of many people from Caltech from their usual research to war work, teaching of undergraduates of course continued. But Carl Anderson (who discovered the positron in 1933) and I were highly critical of the lower-division textbooks that had been used for some time and which had been written by Millikan, Roller, and Watson. So we decided to totally rewrite the volume dealing with electricity and magnetism. It was a crash effort and resulted in a new mimeographed textbook.[6] We retained the names of the authors of the book that this work replaced, although, with the exception of Watson, it had been impossible to consult them. Our new book was used for years, but was never published fully in print. It did, however, contribute to my continuing preoccupation with the teaching of the fundamentals of electricity and magnetism.

The year 1942 was a banner year. I received my Ph.D. from Caltech, and then was given a National Defense appointment as associate physicist. Then I received my U.S. citizenship after the president of Caltech and other members of the senior staff sent appeals to the Immigration Service pointing out that my citizenship would assist the defense effort. And then I got married to DuMond's eldest daughter, Adele. After graduating from high school, she had been accepted at the UCLA campus, but after we made our decision to marry, she transferred her studies to Pasadena Junior College. Because her parents had some family complications, we decided to get married secretly before a Justice of the Peace. Unfortunately, our filing for intent to marry got published in the newspapers, and various people congratulated Adele's parents, to their astonishment. Nevertheless, all calmed down and Adele and I escaped for a few days from our work at Caltech. Incidentally, some weeks later, the Justice of the Peace was impeached—not for marrying us—but for embezzling money. As far as I know, 64 years later, the marriage is still legally valid.

In addition to continuing at the junior college, Adele did a wartime job computing for the Caltech aeronautics department using the then-prevalent mechanical computers to analyze wing flutter on military airplanes. Those computers were large, noisy, and slow. I recall engaging in a race with my fellow graduate student Luke Yuan, later a prominent scientist at Brookhaven National Laboratory. I used a mechanical computer, and Luke used a Chinese abacus. He won.

On the occasion of Adele's 20th birthday, we took a brief vacation, driving to the beach at Santa Monica. Adele was expecting, but still two months from the "due date." We walked quite a bit, and then embarked on the way home. Adele said she was "feeling peculiar." We stopped at a Catholic hospital in Santa Monica, and she had twins in 15 minutes. I never would have made it home. The babies were born prematurely; because our vacation was on the occasion of Adele's birthday, the day of birth of the twins and her own differ by only two days. I started my life as a father by engaging in a scientific argument with the obstetrician who attended the delivery. He claimed that the twins, a boy and a girl, were "identical," that is, originating from the same ovum. Although my knowledge of biology was scanty, I was dubious about this medical opinion. Therefore, I spent time while staying in Santa Monica to research this matter in the library, and then confronted the obstetrician with the biological truth. During those days, newly born babies and their mother were kept in the hospital for a protracted period of time, so I took Adele home after one week and the babies followed after nearly one month in the hospital. During wartime, it was illegal to send personal telegrams; they were allowed only for business. So, my parents sent me a telegram that read, "Gratified by increase in production; we hope quality matches quantity."

We resumed our life in a small apartment in Pasadena while I continued my largely military work. The situation is best described by a line in a letter I wrote to my parents: "Our life is occupied by sleep, work, and babies. About work I am not allowed to write, about sleep I know nothing, and therefore this letter will deal with babies."

The work on the Firing Error Indicator took an interesting turn. Luis W. Alvarez, the highly inventive and resourceful physicist, was working for the Manhattan Project at Los Alamos, and was asked in 1944 by J. Robert Oppenheimer to devise means of measuring the yield of nuclear explosions, both during the first test of the plutonium device and subsequently during the actual use of nuclear weapons against Japan. Alvarez had read some of our reports on measuring shockwaves from supersonic bullets, and decided that our approach might satisfy the assignment he had received from Oppenheimer. Alvarez was one of those people who had the opposite of what is generally known as the "not invented here" syndrome. This implies that if someone else had solved the problem which he was undertaking, he would rather have that someone else complete the work. Accordingly, Alvarez approached Jesse DuMond during the time I was conducting tests of the FEI device on the East Coast. The three of us got together after Alvarez got clearances for the Manhattan Project both for DuMond and me. As a result, I became a "consultant" to the Manhattan Project, and began commuting between Pasadena and Los Alamos.

Much has been written about the history of the Manhattan Project in general, and Los Alamos in particular,[7] so I recite only my experiences. No one else from the FEI Project, including DuMond, participated in the Manhattan District work. I would take the train to Lamy, New Mexico, a small settlement on the Santa Fe Railroad. I would be met there and be driven first to Santa Fe, where my visit to "The Hill" would be arranged. I would then proceed to Los Alamos by car.

It was relatively straightforward to adapt the Firing Error Indicator shockwave detector with its condenser microphones to the application of measuring shockwaves from nuclear explosions. That use only required one microphone, so only one hemisphere carrying the microphone was mounted on a large battery case. The capacity change of the microphone directly modulated the frequency of the transmitter, therefore it was easy to design an automatic calibration device that changed the static pressure in the microphone chamber by a predetermined amount, resulting in a measurable frequency shift. In addition, a further stage of RF amplification was added to extend the transmission range.

When on location in Los Alamos, I was invited to attend all seminars discussing progress on the atomic bomb. J. R. Oppenheimer had insisted that there should be no compartmentalization of information among the Los Alamos scientists; I was one of them, although I had no experience or expected involvement in bomb design. Thus I received a limited education on the subject; however, because of the urgency of designing and building the shockwave detector, I had very little time to become more involved. I participated in what became known as the RaLa (radioactive lanthanum) Experiment. This isotope was used as an embedded source to examine the time sequence of the implosion in the plutonium device by observing the intensity of penetrating gamma rays as a function of time. I recall sitting on a cliff after the explosion dispersing the radio-lanthanum, waiting for clearance from the radiological officer to get off the cliff and come down to the ground without excessive radiation exposure. It became dark, and I urged the officer (Dr. Louis Hempelman) to let me come down before it became too dark to descend safely. The answer I received was, "It is my responsibility to protect you from excessive radiation, not to keep you from breaking your neck." I climbed down without authority.

We worked extremely hard getting the shock wave detection device ready for the Trinity test of July 16, 1945. Alvarez, Larry Johnston (now at the University of Idaho), Bernard Waldman (then at Notre Dame), and Harold Agnew (who later became the director of Los Alamos), and I all boarded a B-29 airplane designed to deploy the shockwave gauge by parachute. However, because of the bad weather, and the (at that time) uncertain explosive power of the Trinity device, Robert Oppenheimer ordered Luis Alvarez not to approach the test location closer than 25,000 feet, an order that Alvarez had to accept despite vigorous protest. We took off as planned, and were not able to deploy the device at that distance, but made extensive sketches of the mushroom cloud (as is described in detail in Alvarez' autobiography[8]). At that moment I was much too exhausted from getting ready for the test to worry about the profound implications of being a participant in ushering in the nuclear weapons age, but that concern deepened over time.

During my participation in Los Alamos, I found that the majority of the scientists were not preoccupied with the historical implications of their work. As is well known, a petition was generated at the University of Chicago asking that, before military use, a bomb be exploded in a demonstration to which Japanese observers would be invited. Town meetings were held, organized by Robert R. Wilson, to consider such questions at Los Alamos. This matter was considered by "the Interim Committee" advising the Secretary of Defense to which Oppenheimer, among other physicists, was an advisor. It was decided that a demonstration was unlikely to be effective, and might harden the will of the Japanese rather than accelerate their surrender. In the words of the scientists, "We can propose no technical demonstration likely to bring an end to the war; we see no acceptable alternative to direct military use."

The bombing of Hiroshima and Nagasaki proceeded, and the shockwave gauges were dropped by parachute from a plane separate from the one dropping the bomb; Alvarez, Johnston, Waldman, and Agnew participated in that mission. The shockwave signatures were transmitted and photographed on an oscilloscope aboard the plane that had released the parachute. Analysis of these shockwave images received over Hiroshima induced a downward revision of the estimated yield from 20 kilotons to 13 kilotons. It is noteworthy that a shockwave from an explosion has the form of the letter N. This means that there is first a compressional steep wavefront, steepened by the fact that the speed of sound increases with pressure and therefore the crest of the wave catches up with the lower pressure of the initial disturbance. Following that steep wavefront is a rarefaction, which is then sharpened, resulting in a very steep diminution of pressure. Both the amplitude of the shockwave, as well as the time interval between the rarefaction and compression, provide an independent measurement of the strength of the explosion.

One interesting sideline of the use of these devices is worth noting here. Alvarez decided to attach a letter to the battery case of our shockwave measuring device when it was dropped by parachute over Nagasaki. The letter was addressed to Ryukichi Sagane, a Japanese physicist who had worked at the University of California at Berkeley on beta ray spectroscopy before the war and who had been specifically encouraged to prepare to copy the UCRL 60-inch cyclotron in Japan. This letter explained the general nature of the nuclear explosion, and asked Sagane to explain the futility of further resistance to the Japanese High Command. The letter was picked up intact by Japanese Naval Intelligence and rapidly delivered to the Japanese High Command. It is not known historically, at least not to me, whether that letter had any influence on the Japanese decision to accept President Truman's demand for unconditional surrender. Much later, after the war, a copy of that letter which Alvarez had composed and which was edited by Robert Serber and Phillip Morrison was officially signed by Alvarez and presented to Sagane during a postwar visit to the University of California Radiation Laboratory.

There has been and continues to be vigorous controversy over President Truman's decision to use the bomb against Japan well after Germany had surrendered on May 8, 1945, and also about the decision to use a second bomb over Nagasaki.[9] It was well known that the Japanese were very seriously considering surrendering at that time, although there was strong resistance on their part to Truman's demand for unconditional surrender. Truman, who became President on April 12, 1945, was presented with estimates of expected casualties, both Japanese and American, should it be found necessary to invade Japan, an invasion then scheduled for November 1, 1945. Some of these estimates made public were greatly excessive.

Many more people had died as a result of World War II (45 million altogether) than the approximate one-quarter million Japanese who died at Hiroshima and Nagasaki as a result of the combined immediate and delayed lethal effects of the bomb. Civilians had been targeted before, both by the Allies and the Axis powers. Neither Roosevelt nor Truman had ever previously contemplated not using the bomb after the enormous effort to produce it.

Alvarez[8] absolutely self-righteously claimed that all American actions were justified. However, many historical accounts of the events, and particularly those of Barton Bernstein[10] of Stanford University, have cast doubts on the validity of the estimates of casualties that Truman was given, and even on whether the dropping of the second (Nagasaki) bomb was ever explicitly authorized by the President.

Clearly the prospect of Soviet entry into the war and the shaping of the postwar United States–Soviet relationship played a major role in the timing of the Hiroshima and Nagasaki attacks. At the Potsdam Conference on July 24, 1945, Truman stated casually that the United States had developed "a new weapon of unusual destructive force," without identifying the physical nature of the weapon. The extent of the Soviet program in nuclear weapons technology and the successes of Soviet espionage were not then known. The highly complex nature of the "Soviet factor" in the U.S. decision has been extensively discussed in the literature.[11] I cannot judge the merits of all these controversial factors then in play, but I believe that the most important questions raised by these horrendous events concern their impact on the future accumulation and potential use of nuclear weapons, and the proliferation of nuclear weapons technology all over the globe. These future risks outweigh in importance the shorter-range considerations that controlled the decision to drop the bomb. These nuclear dangers have by no means disappeared today.

During my collaboration at Los Alamos with Luis Alvarez, he speculated extensively about using surplus radar equipment remaining from the war—operating at a wavelength near 200 MHz—as radio-frequency sources to power linear accelerators to be built once he returned to the University of California Radiation Laboratory (UCRL) at Berkeley. According to E. Lofgren,[12] Alvarez first intended to design an electron machine, but his interest shifted quickly to protons. At that time, linear accelerators had not been used in nuclear science, although a preliminary version of a heavy ion linear accelerator of low energy had been built by Sloan and Lawrence. Alvarez hoped to use the availability of "cheap" radio-frequency sources to have linear accelerators reach high proton energy, eventually exceeding the energy attainable by cyclotrons. The argument was that presumably the cost of linear accelerators would scale linearly with energy, whereas the cost of the then-existing cyclotrons, which had been the only source of high-energy particles beyond those provided by electrostatic machines, would scale as the third power of the energy.

4
Work at the University of California Radiation Laboratory

After the war ended, I had to make decisions about my future work. Alvarez strongly urged me to go back with him to join the staff at the University of California Radiation Laboratory (UCRL) and help him translate his ideas about proton linear accelerators into practice. I also had an offer for an industrial position at what was then the Bell Telephone Laboratories, and there was a strongly expressed interest in having me join the faculty at my alma mater, Princeton University. I decided to stay in California, and follow Luis Alvarez's lead. I had no experience whatsoever in nuclear physics, high-energy particle physics, or the design and operation of high-energy accelerators, nor even in radio-frequency or microwave technology, but the idea proposed by Luis Alvarez was enormously appealing to me, so I accepted a staff position at UCRL.

We moved from Pasadena to the Bay Area driving "Old Faithful," our beaten-up 1937 Dodge sedan. Finding housing in Berkeley proved impossible at any reasonable cost, so we rented a small house in Concord, California, about one hour's driving distance from the Berkeley laboratory. It was a nice little place with a yard for the twins to play outdoors. Happily, there were several others in the same position. A close neighbor was Herschel Snodgrass with his wife Betty; Herschel was an instructor at the Berkeley physics department, deeply interested in improving teaching of science. Also close by was V. L. VanderHoof, generally known as "Van." He was a Stanford professor of geology on leave, and an expert paleontologist. However, during the war, he had decided to dedicate his considerable mechanical skill to war work by becoming a technician at UCRL. We became friends with both the Snodgrasses and the VanderHoofs, and there was plenty of opportunity for ride sharing.

The Radiation Laboratory was started on the UC campus in 1931 by E. O. Lawrence after he had invented the cyclotron in 1928. As the Laboratory grew, most of it moved "up the hill" behind the main campus. UCRL accommodated cyclotrons of increasing size; the largest one was planned as a conventional cyclotron to reach an energy of perhaps 100 MeV deuterons. To achieve such large energy and to avoid the lack of synchronism produced by the relativistic mass increase of the particle to be accelerated, such acceleration had to be achieved in very few turns, and this required very high voltages—above a million volts—on the Ds of the cyclotron which were fed by a radiofrequency oscillator of large power. This high voltage, in turn, required a large magnet gap, which limited the strength of the magnetic field. During the war, the plans for such a giant conventional cyclotron were suspended and the magnet was instead dedicated to a massive program to separate uranium isotopes using electromagnetic isotope separators called Calutrons. The developmental work at UCRL laid the basis for the Y-12 electromagnetic isotope separation plant at Oak Ridge, Tennessee that supplied the Uranium 235 for the Hiroshima bomb. All this is well documented elsewhere.

After the war ended, Lawrence intended to reconvert the 184-inch magnet back to the originally planned cyclotron use. When I first arrived at UCRL and entered the hall where the giant magnet was located, I was greeted with a loud shout of, "Hey, we need you!" This originated from Duane Sewell, then in charge of magnetic measurements.[1] It turns out I was the only person within range of Sewell's view who was short enough to stand up inside the poles of the magnet, which had a gap of five feet. Thus I spent the first two days of my employment at UCRL incarcerated inside the vacuum chamber of the 184-inch magnet making magnetic measurements. I mean "incarcerated" literally, because the side plates of the vacuum chamber had to be in place because they, being made of iron, perturbed the magnetic field.

Very soon thereafter, it was decided to convert the 184-inch installation from a conventional cyclotron to a synchrocyclotron of much smaller gap, with the Ds excited by an oscillator operating at a frequency modulated by a rotating condenser. Therefore, the magnetic measurements had to be repeated for the smaller gap, without the presence of a human being inside that gap.

The UC Radiation Laboratory pioneered what is now generally designated as "Big Science." Lawrence, as director, would make unquestioned decisions. Before the war, Lawrence had been a very successful fundraiser from private sources and foundations; after the war, partially in recognition of his leadership of the electromagnetic separation project, the government supported his laboratory liberally. The organization of the laboratory was almost unstructured. Senior individuals gathered capable people around them to execute the various programs. I functioned more or less as what today would be called "project manager" for the construction of the Proton Linear Accelerator that Alvarez had persuaded Lawrence to support. To illustrate the relative informality of those days, I might mention that I never knew the budget for this project.

In the past, most large accelerators have been poorly documented. In particular, there exists no complete report on the construction and performance of the family of cyclotrons constructed during the history of UCRL. I felt very strongly that this should be changed, and after completion of the linear accelerator, an extensive article was prepared.[2] Further details on the history and technical facts covering proton (and electron) linear accelerators are given in the monumental volume edited by Lapostolle and Septier.[3] In the material that follows I therefore only discuss some of the episodes in which I was involved, rather than giving an overall account of the history of the proton linac.

During my time as a staff member of UCRL, that laboratory continued to carry out some classified work. Most of this was in the area of nuclear chemistry, particularly investigations of the transuranic elements. As a result, all members of the staff wore badges differentiated by their level of access. Moreover, E. O. Lawrence and Luis Alvarez remained involved in the nuclear weapons programs and some staff members later participated in the nuclear weapons test series in the Bikini Atoll. Initially, after arriving at Berkeley, I was anxious to disseminate facts on the revolutionary impact of nuclear weapons on the security future of the country. Accordingly, I gave talks to a variety of lay groups largely designed to describe the vast difference in energy released in nuclear processes as compared to that released in chemical reactions. I talked to groups such as labor unions, women's clubs, and service organizations. Most of this effort proved frustrating and unproductive. (I remember after a talk to a labor union, someone asked me, "What are you, some kind of Commie?") Thereafter, I remained uninvolved in these issues for some time, in part because the accelerator and research work at UCRL during that period became truly exciting and productive.

The group gathered by Alvarez assumed diverse responsibilities in an informal way. The linear accelerator, designed to attain a proton energy of 32 MeV, used a 4-MeV electrostatic generator as an injector, patterned after the machines at the University of Wisconsin. Alvarez had hired Clarence Turner to be in charge of building that machine under his own independent direction. Note that 4 MeV was a record energy for such machines at that time. The main accelerator was designed to be 40 feet long and contained a radio-frequency structure excited as a single cavity. That cavity was fabricated at the Douglas Aircraft factory with which Luis Alvarez had previously worked.

The original intent was to excite this cavity with a number of the surplus radar oscillators that were part of the "268" early warning radar system that had become obsolete after the war. Essentially, the main process of development consisted of throwing away all the surplus components that had given the main impetus to justifying this accelerator to start with, and replacing them with components properly designed for their function. The surplus oscillators were replaced by redesigned units using commercial radio transmitting tubes. The separate modulators feeding each oscillator were replaced by a single central installation.

A schematic diagram of the cavity contained in the main vacuum tank of the accelerator is shown in Figure 4.1. That cavity was to operate in the lowest TM_{01} mode, giving essentially uniform acceleration for its entire length. The frequency of each oscillator was controlled directly by this cavity through "tight coupling" through coaxial feeds. To convert this schematic concept into a working machine, an enormous amount of work had to be done. A group,

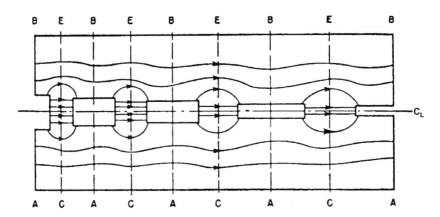

FIGURE 4.1. Electric field lines in the "Alvarez Structure" proton linear accelerator. (From L. Alvarez et al.[4])

led by J. R. Woodyard, who had extensive microwave experience from his work at Stanford and the Sperry Gyroscope Company, and by Frank Oppenheimer (brother of J. Robert Oppenheimer), conducted extensive measurements on model cavities at a scaled-up frequency of 3000 MHz. A separate group at the Laboratory, led largely by William Baker, developed the new oscillators that replaced the surplus units. In addition to handling numerous technical problems, I worked extensively on some theoretical analyses associated with the machine, notably on orbit dynamics[5] and on control of the electromagnetic field in the cavity.[6]

The principle of phase stability, independently discovered by E. M. McMillan and Vladimir Veksler, became widely accepted and, as noted above, led to the total redesign of the 184-inch cyclotron. That principle could also be applied to the phase stability of the proton bunches accelerated in the proton linear accelerator. Ed McMillan[7] had shown that radial focusing and phase stability in a linear accelerator were mutually incompatible unless either (a) charge was included in the beam, (b) external magnetic or electrostatic focusing devices were used, or (c) the time-varying character of the field was sufficient to produce focusing.

We chose the first alternative, which is introducing charge included in the beam induced in a metal conductor. Hugh Bradner succeeded in manufacturing extremely thin beryllium foils that would cover the entrance aperture of each drift tube. The problem was that as soon as these foils were installed and the machine was turned on, they all disappeared, pulled out by the electrostatic forces. After this initial setback, the foils were replaced by grids of sufficient openness to intercept only a tiny fraction of the beam.

I frequently presented beam dynamics calculations during the regular UCRL seminars during which progress of various activities was presented.

These were well-attended events with E. O. Lawrence sitting in a red upholstered easy chair reserved for his own use. My talks were sufficiently dull that I believe he rarely exhibited any interest. I also gave quite a few other talks on recent developments in microwave technology and associated fields.

The problem of controlling the field distribution in the 40-foot long accelerating cavity proved to be more difficult than anticipated. In essence, the cavity structure can be envisaged as a series of subcavities (Figure 4.2), and these were the units exhaustively modeled by the microwave group. However, because each successive cell had to increase in length to match the increasing velocity of the proton, the electromagnetic fields of successive sections did not exactly match, and therefore, the field distribution in the entire cavity did not actually produce the uniformity desired. We showed theoretically that the sensitivity of field distortion in such a long cavity varies quadratically with the ratio of the cavity length to the wavelength. This follows from the fact that the separation of the next resonant mode from the lowest mode which we tried to attain also decreases quadratically; the two lowest modes "mix" and distort the distribution. I was able to develop a procedure to convert a Fourier analysis of the observed longitudinal distribution of the accelerating fields into a correction procedure executed through adjusting the length of each drift tube.

Another problem observed was that of multipactoring. Minor discharges of x-rays would eject secondary electrons from the wall which, in turn, would travel in the electromagnetic field of the cavity, hitting the wall again, leading

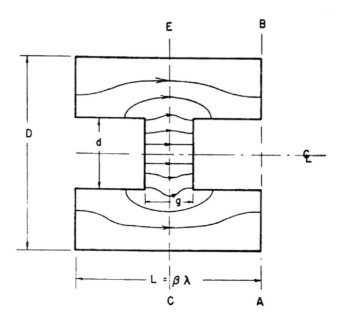

FIGURE 4.2. A unit cell in the linear accelerator cavity. (From L. Alvarez et al.[8])

to an exponential buildup. This phenomenon was initially controlled by applying electrostatic voltage to the drift tubes; but after a while we learned to control the phenomenon instead by a sufficiently rapid buildup in the fields in the cavity.

The foregoing is only a brief outline of some of the problems on the project in whose solution I directly participated. Our work was interrupted by one major disaster. The electromagnetic cavity was raised from and lowered into the vacuum enclosure by two cranes. The mechanical engineering department designed a gadget for one of the cranes that permitted slow-motion raising or lowering of the cavity. One day this device jammed the switch of the crane in the "on" position, with the result that the chain lifting the hook was torn apart and the load dropped, smashing one end of the cavity. Alvarez was able to negotiate a very rapid replacement of the cavity, and no one was hurt by the drop.

Overall, design and construction of this accelerator was a highly successful enterprise thanks to all those helping with the endeavor. However, the extension of proton linear accelerators to an energy of several GeV never became reality because the invention of the proton synchrotron, again flowing from the principle of phase stability, converted the scaling law of such machines to one also roughly projecting a linear cost increase with energy. Thus the proton linac never caught up. Our machine was used for particle physics research and then transferred to the University of Southern California for further research. A somewhat longer, similar machine was constructed at the University of Minnesota and today, nearly all injectors into proton synchrotrons are linear accelerators of the drift tube type as developed at UCRL, but generally incorporating electromagnetic focusing. The 800-MeV proton accelerator (LAMPF) built at Los Alamos and the 1 GeV machine at the Spallation Neutron Source at Oak Ridge are the largest proton linear accelerators constructed to date.

At UCRL there was no distinction between particle physicists and accelerator physicists. It was generally expected that most experimental physicists would work on accelerators and then as a "reward" for their successful contributions to the machines, they would have opportunities to do experimental research. There were no organized particle research groups. After the linac commenced operation, I first worked on an experiment on proton–proton scattering at 32 MeV using the proton beam of the linac. Interestingly enough, this required extending the proton beam pipe beyond the confines of the building housing the linac itself, through one of the stalls and out the other side of the men's room, leading of course to the inevitable jokes. The experimental apparatus was then constructed outside the men's room, which remained operational.

Before proceeding to outlining my particle research activities using both the newly constructed 32-MeV linac and the other accelerators built at UCRL, let me digress to some other matters relating to life at Berkeley. One of the many positive aspects of work at UCRL was the presence of Robert Serber, the theoretical physicist, who had played a major role at Los Alamos.

He was a theoretician communicating extensively with experimentalists. In addition to advising on orbit dynamics and the future program of the Laboratory in particle physics, he gave a series of excellent lectures that were distributed but not formally published under the title of "Serber Says." These lectures gave me a most valuable introduction to the current state of particle physics.

There was also extensive social life at UCRL in which Serber played a large role. Lawrence hosted the staff on the occasion of many "celebratory milestones." I remember a party to honor Hideki Yukawa, the Nobel Prize-winning Japanese physicist, who attended with his wife. We were dancing, or in my case, hopping around, which is all my limited skill permitted. During intermission, I sat next to Yukawa, who watched his wife enjoying herself on the dance floor. He turned to me and said, "Do you believe, Panofsky, that after this visit, my wife will still be obedient?" I don't recall my reply. Serber was a close friend of Frank Oppenheimer, who was part of the team working on the 32-MeV linac. We became good friends of both the Serbers and the Oppenheimers; the latter had children close in age to our twins.

In 1946, my staff position was changed to assistant professor at the university. I recall that Professor R. T. Birge, with whom I had communicated previously in connection with work on natural constants, called me to his office, and said, "Panofsky, do you have time Monday, Wednesday, and Friday at 8 AM?" I had been forewarned that Birge was going to offer me an academic appointment, so I replied in the affirmative; then he made the formal offer and I accepted. This meant that I had to complement my work at UCRL with teaching on campus. This new appointment also made it possible for us to move our family from Concord to the Claremont district of Berkeley where we bought a rambling old house quite near to UCRL. We thus joined the American dream of home ownership and a large mortgage.

Our family increased in the spring of 1947 with the addition of a son, Edward Frank, to the twins Richard and Margaret, who were now approaching kindergarten age. Edward was born at the Kaiser Hospital in Oakland, California. The University offered me Kaiser membership quite soon after my arrival in Berkeley, and I am still a member with a medical record number starting with four zeros, making me almost a founding member of that enormous HMO.

My teaching duties were fairly extensive. I taught a graduate course in electricity and magnetism and designed apparatus for an upper-division laboratory. During my stay at Berkeley, I supervised the Ph.D. work of fourteen graduate students. The teaching of electricity and magnetism was challenging inasmuch as there were no textbooks I thought were suitable at that level. I was trying to teach a course intermediate between that taught by Smythe at Caltech, with its heavy emphasis on problem-solving, and a purely theoretical approach. My goal was to emphasize the experimental origins of each of Maxwell's equations rather than their theoretical structure as such. I also hoped to include the classical theory of the electron. I attempted to secure permission to translate the German books of Abraham and co-authors but at the time this required permission from something called the Foreign Property Administrator, which would incur large delays. I therefore prepared a mimeographed text of my own, which went through several editions benefiting from the constructive and destructive comments of my graduate students. I was then advised that it would be good to convert these notes into a formal textbook, but I also recognized that I simply did not have time to do that without a collaborator, considering all the work I was also undertaking at UCRL.

While visiting, Professor E. U. Condon informed me that Dr. Melba Phillips had just been dismissed from her teaching position at Brooklyn College as a result of refusing to identify her colleagues during hearings before the U.S. Senate Internal Security Subcommittee. Condon suggested that she might be interested in becoming my collaborator. I contacted her and she accepted, although we had never met. The book[9] was produced entirely by mail correspondence between us: I produced drafts by dictating to my secretary Velma Turner, and then sent this material to Melba Phillips. I later met her in person at Condon's house, and we remained friends throughout her long lifetime. She was a wonderful woman who, among other honors, served as president of the American Association of Physics Teachers. I joined as a co-author writing her obituary for *Physics Today* in 2005.[10] The textbook we wrote had a wide distribution and was translated into Russian, Hindi, Italian, and Japanese. It went out of print for a while, but was recently republished.

Let me return to my research work at UCRL. My first elementary particle experiment was proton–proton scattering at 32 MeV.[11] We decided to examine proton–proton scattering by two alternate techniques. I chose to design a precision camera in which proton-track-sensitive emulsions were mounted. In parallel, Bruce Cork, Lawrence Johnston, and Chaim Richman[12] carried out an experiment using electronic detection methods. Previously, a series of experiments of p–p scattering had been carried out up to 14.5 MeV and showed no deviation from pure S wave scattering.

Theory based on the then-prevalent concept of nuclear forces predicted that some P wave and D wave contributions should become manifest at the higher energy of 32 MeV. Our experiment involved the scanning of over 10,000 tracks, and was carried out in the demonstrated absence of significant background. Figure 4.3 shows our results plotted together with those of Cork et al. The most significant result was that, even at this higher energy, contrary to theoretical expectations, no contributions from the higher angular momentum partial waves were evident. The result gave some indication, much discussed by theorists at that time, of an accidental cancellation among the higher partial wave amplitudes. I gave a seminar on the results at Princeton University. To my surprise, Wolfgang Pauli was present and seemed to be highly interested, signaling his attention through the nodding of his head at high frequency. I also participated in a small experiment[12] on the cross-section for producing the radioactive isotope C_{11}.

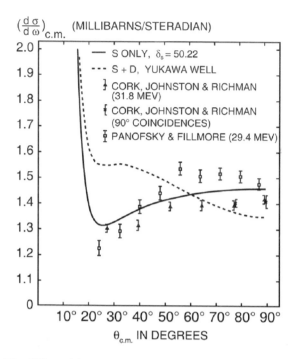

FIGURE 4.3. The differential scattering cross-section for proton–proton scattering at 32 MeV as a function of center-of-mass scattering angle. (From W. K. H. Panofsky and F. Fillmore.[14])

The work on the linac was interrupted by an explosion of one of the nitrogen cylinders which stored that gas outside the accelerator building for use in pressurizing the electrostatic generator injector. The cylinders were wartime surplus, and some of them were rolled from steel in two layers. If the inner layer developed a small leak, then the stress on the outer layer would double, thus constituting a veritable time bomb. I received a phone call at eight o'clock one evening telling me that an explosion had destroyed a wall of the accelerator building.

I went to the laboratory and found that R. F. Mozley, whose thesis experiment was located in the building, had signed in to the Laboratory but had not signed out and had not been observed to leave. Mozley, a graduate student of Luis Alvarez, was doing a very ingenious experiment designed by Luis to measure the half-life of the neutron. Neutrons emerging from a target in the electrostatic generator were to traverse the walls of a circular storage ring vacuum chamber that would capture some of the decay electrons from the passing neutrons. I saw the rubble of Mozley's experiment and searched for his body. After failing to find him, I drove to his home and, to my great relief, found him painting the wall of his bathroom. So much for entrance and exit control by the security people. Mozley changed his thesis topic and instead did a photon experiment directed by Jack Steinberger.[15] It was fortunate indeed that the explosion occurred "after hours"; broken glass was projected all over the nearby unoccupied cafeteria.

Having earned my "spurs" on the 32-MeV linear accelerator, I was also encouraged to start using the other machines which by that time had sprouted at UCRL. Ed McMillan enlisted me first to map the external neutron beam of the 184-inch cyclotron by activating some materials and analyzing their radioactivity. Then followed the highly exciting period at UCRL when the 184-inch cyclotron succeeded in artificially producing pi mesons or pions. Charged pions had been discovered previously in the cosmic radiation. The positive identification of charged pion production in the laboratory was made by Gardner and Lattes;[16] the latter had been recruited as a visitor to UCRL after his participation in the cosmic ray discovery. The work was done with photographic plates, and succeeded after Lattes taught his collaborators the fine art of processing the photographic plates so that weakly ionizing tracks could be detected.

Initially, the 184-inch cyclotron was operated unshielded, despite the very substantial neutron fluxes. Lawrence decided to correct this deficiency, and in an amazingly short time, the Laboratory procured large quantities of massive shielding blocks surrounding the accelerator.[17] This made it possible for a collaboration led by Burton Moyer to set up a detector for gamma rays emanating from an internal target of the cyclotron and then passing through a hole in the shielding.[18] That work strongly hinted that neutral pions decaying into a pair of γ-rays were being produced at the internal cyclotron target; this had been predicted theoretically and had also been speculated upon from observation of gamma ray fluxes in cosmic rays.

The photographic plate experiments by Lattes, Gardner, and collaborators conclusively demonstrated the phenomenology of charged negative and positive pions. Positive pions exhibited their decay into muons, which were presumed to be identical to the penetrating components of cosmic rays.

Negative pions, when stopped in the emulsions, produced clearly visible nuclear disintegrations. By momentum conservation negative pions, if stopping in hydrogen, could not convert the proton into a neutron without emission of an additional, presumably neutral particle. I therefore devised an experiment placing a high-pressure hydrogen vessel with thin walls in close proximity to the internal cyclotron target and then aligning a gamma-ray pair production spectrometer, similar to the one designed by Moyer and collaborators, with an aperture in the shielding in line with the hydrogen vessel. Using hydrogen compounds such as lithium hydride or CH_2 (polyethylene) would not have worked because the higher atomic number components would absorb the pions highly preferentially. The experimental arrangement is shown in Figure 4.4. Calculations indicated that with this arrangement, observation of gamma rays would be possible at measurable rates if the reaction occurring were

$$\pi^- + p \rightarrow n + \gamma. \tag{4.1}$$

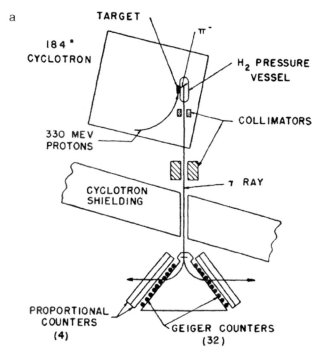

FIGURE 4.4a. Physical arrangement for the experiment to measure the γ-ray spectrum from absorption of negative pions in hydrogen. (From W. K. H. Panofsky, R. L. Aamodt, and J. Hadley.[19])

Two graduate students, Lee Aamodt and later Jim Hadley, joined the experiment. The spectacular results were published, first in preliminary form[20] and then in final form,[21] after having been discussed at various conferences.

We designed a pair spectrometer in which the incident gamma rays were converted into electron–positron pairs inside a magnet that would then deflect them into counter arrays on both edges of the magnet. Sixteen Geiger counters detected the electrons and positrons. These counters provided sufficiently narrow channels and were backed by proportional counters of better time resolution. Counting rates were exceedingly slow, so we simply observed the arrival of the electrons and positrons by light flashes from small neon tubes. Whenever two neon tubes flashed in apparent visual coincidence we would throw an iron washer over the appropriate nail in a square array of nails (Figure 4.5). By watching the accumulation of washers along a diagonal of this array, we could observe the growth of gamma rays of a specific energy. A very primitive coincidence circuit indeed! Later a paper recording system was employed.

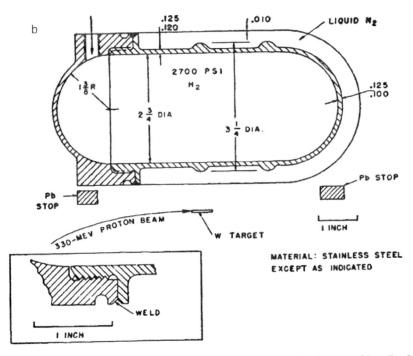

FIGURE 4.4b. The hydrogen pressure vessel. (From W. K. H. Panofsky, R. L. Aamodt, and J. Hadley.[22])

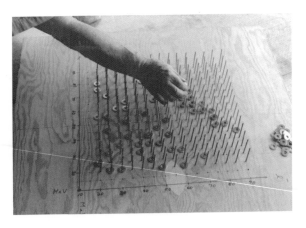

FIGURE 4.5. The "biomechanical" coincidental spectrum. (From the Panofsky Collection.)

The results proved extremely exciting. There emerged two groups of gamma rays separated in energy: first, the expected peak from reaction (4.1) cited above, but then there also appeared a broad gamma-ray distribution at lower energy which we plausibly interpreted as likely originating from the reaction

$$\pi^- + p \rightarrow n + \pi^\circ, \tag{4.2}$$

with the π° decaying into two γ-rays. The spectrum is shown in Figure 4.6.

The interpretation of this experiment bore fruit in many directions. The first reaction (reaction 4.1) gave a measurement of the charged pion mass with a precision comparable to, or exceeding, that previously available. In addition, that reaction provided guidance to calculating the coupling strength of pions to the nucleon in comparison to the electromagnetic force. Reaction (4.2) confirmed the existence of the neutral pion, assuming it to disintegrate into gamma-ray pairs. It also showed that the neutral pion was lighter than the charged pion and provided a measurement of the mass difference. The masses of both pions were determined with an accuracy of close to one percent. At that time, it appeared indeed fortuitous that the reaction rates of processes (4.1) and (4.2) were comparable so that both could become evident in a single measurement, but phase space calculations made this result reasonable. Also, reaction (4.2) showed that neutral and charged pions have identical spin-parity properties.

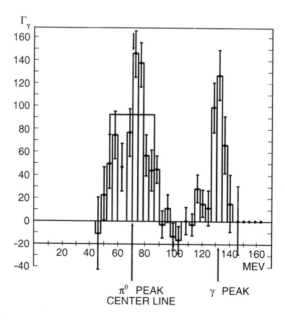

FIGURE 4.6. The γ-ray spectrum from the absorption of negative pions in hydrogen. (From W. K. H. Panofsky, R. L. Aamodt, and J. Hadley.[23])

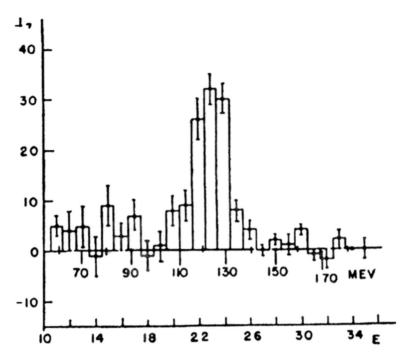

FIGURE 4.7. The γ-ray spectrum from the absorption of negative pions in deuterium. (From W. K. H. Panofsky, R. L. Aamodt, and J. Hadley.[24])

An obvious sequel to examining the gamma rays from absorption of negative pions at rest in hydrogen was the examination of absorption in deuterium. Although the apparatus remained essentially the same, provisions had to be made for the recovery of deuterium gas. In this case, the gamma-ray spectrum (Figure 4.7) again showed the spectrum from the process,

$$\pi^- + d \rightarrow n + n + \gamma \qquad (4.3)$$

but the emission of a broad peak analogous to reaction (4.2); that is,

$$\pi^- + d \rightarrow n + n + \pi^\circ \qquad (4.4)$$

was not observable. However, the absolute reaction rate observed was such that the rate of

$$\pi^- + d \rightarrow n + n \qquad (4.5)$$

could be inferred with confidence. In fact, about 70% of the total rate appeared to originate from reaction (4.5) and 30% from reaction (4.3). From this latter observation, the important conclusion could be drawn that the negative pion was a pseudoscalar particle, that is, that it most likely had spin zero and negative intrinsic parity. To draw this conclusion, it was necessary to assume that the negative pion was captured from its lowest electromagnetic orbit around the deuteron, an s state of zero angular momentum.

Theoretical analysis on this point by A. S. Wightman[25] of Princeton University showed that the lifetime of a π^- captured in an orbit of high angular momentum or high principal quantum number to return to the ground state would be much shorter than the decay time of the negative pion; therefore capture on the nucleus would take place from the orbital ground state. The nucleon ground state of the deuteron was known to be a ^3S state of spin 1, whereas from the Pauli exclusion principle, two neutrons must either be in a ^1S or ^3P state. Therefore, the predominance of reaction (4.5) showed that the capture process of the π^- in reaction (4.5) had to involve a change of parity, thus leading to the conclusion that the negative pion must be a pseudoscalar. Additionally, the rate of reaction (4.5) gave some evidence of the n–n interaction which previously had not been susceptible to observation. At the same time the absence of reaction (4.4) was explained if the π^- and π° had identical parities.

In summary, experiments of π^- absorption at rest gave a plethora of very valuable information on pion properties and other then-open questions. None of this information was contradicted by future developments, but the accuracy of the masses and branching ratios were substantially improved by later experiments, in particular, those at CERN.

In parallel with the experiments on the absorption of negative pions in hydrogen and deuterium, I participated in experiments on Ed McMillan's 300-MeV electron synchrotron which had then started operation. Previously, I had worked a bit with McMillan on getting that machine going by helping solve some problems with its power supply. Once it started operating, I did an experiment,[26] together with two graduate students, measuring the ionization density in a thick multilayered detector produced by the *Bremsstrahlung* (retardation radiation) emitted when an internal target in the synchrotron was struck by high-energy electrons. Such electromagnetic shower propagation was a well-known phenomenon in cosmic rays and had been extensively theoretically computed at very high energies. However, measuring such shower propagation near 300 MeV proved a useful experimental tool.

After these preliminary experiments, I joined Jack Steinberger and his student; they were engaged in measuring γ–γ coincidences from an external target struck by the gamma-ray beam emanating from McMillan's synchrotron. The result of this experiment[27] demonstrated conclusively the decay of the neutral pion into two gammas, and gave an independent measurement of its mass. Figure 4.8 from that paper shows the coincidence rate between the two gammas as a function of the angle between them. Note that the rate shows a sharp cutoff as the angle between the two gamma-ray detectors falls below a certain minimum angle. This cutoff is generated because, in the center of mass system of the neutral pion, the two gammas are emitted at 180°. Therefore, in the laboratory system the angle between the two gammas cannot be any smaller than that generated by the maximum velocity of the neutral

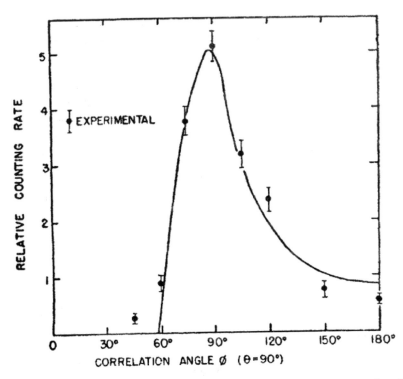

FIGURE 4.8. Coincidence counting rate between the two γ-ray counters as a function of the correlation angle between them, resulting from neutral pion production from an internal synchrotron target. (From W. K. H. Panofsky, J. Steinberger, and J. Steller.[28])

pion as emitted. Steinberger and his collaborators also had measured the photoproduction of charged pions from the synchrotron.[29] The angular distribution of both neutral and charged pions, when compared with theoretical predictions, confirmed the psuedoscalar nature of the pions.

Working with Steinberger was an exciting experience. He was an absolutely dedicated experimenter who worked continuously while the synchrotron was in operation, essentially living off candy bars. Once, Lawrence appeared accompanied by a congressman to whom he wished to show the Laboratory; he asked Steinberger and the synchrotron operators to shut down the machine to show it to his visitor. Steinberger's reaction was, "I'm doing something important," which did not endear him to Lawrence. As we have learned from Steinberger's autobiography,[30] this nonendearment continued, leading to Steinberger's later departure, a great loss to the UCRL.

I returned to the 184-inch cyclotron for further experiments, none of them yielding results quite as exciting as those from the absorption of negative pions in hydrogen and deuterium. With E. Martinelli, a graduate student, we made a measurement[31] of the half-life of the charged pions by installing an internal spiral channel near the internal target of the cyclotron where we could trace the decaying intensity of the pions as they spiraled vertically in the magnetic field. I also worked on extraction systems of beams from the cyclotron. Focusing external beams extracted from circular machines was difficult at that time, because systematic strong focusing had not as yet been discovered, although focusing by magnets of alternating gradients had been known previously.

W. R. Baker and I[32] devised a plasma discharge tube that essentially constituted a transparent current of around 10,000 amperes. When the charged external beam passed along the axis of this discharge, it became subject to first-order magnetic focusing constricting the beam. Because this primitive device could sustain discharges only for a very short time, it was a mismatch to the much longer beam structure of the cyclotron, and therefore was not applied practically at the time. However, similar and more sophisticated devices later came into use for the focusing of external beams as needed for the generation of neutrino beams.

A group of us also repeated the early measurements of Moyer et al.[28] with very high precision. The resulting exact energy spectra of gamma rays produced by an internal cyclotron target corresponded with great precision to that produced by decay of neutral pions into two gammas, not a surprising result after the direct γ–γ coincidence experiments from the synchrotron.[35] A further experiment was conducted by Sue Gray Al-Salam, a graduate student from the South who had married an Iraqi colleague. Her experiment successfully examined the production of fission induced by stopping negative pions in photographic plates loaded with a uranium compound. Indeed, such fissions could be detected. Later, Ms. Al-Salam accompanied her husband to Iraq but found it very difficult as a female Ph.D. physicist to practice her profession there; even when conducting a demonstration experiment in front of students, she was expected to have a male assistant move all apparatus for her.

Buoyed by the success of the research programs on the existing accelerators, Lawrence pushed on to the construction of the "next step" in energy. Throughout the prewar and postwar period there continued to be competition or even outright tension between East and West, both in respect to leadership in accelerator-based particle physics and, ironically, also in nuclear weapons policy. The exponents of this East–West tension were I. I. Rabi and E. O. Lawrence. In respect to accelerators, this tension was resolved by the Atomic Energy Commission (AEC) authorizing UCRL to proceed with the "Bevatron" while the 3-GeV (3 billion electron volts) Cosmotron was constructed in the East at Brookhaven, Long Island. William Brobeck, the highly gifted engineer who had assisted Lawrence throughout most of the existence of UCRL, was the designer and chief engineer of the Bevatron. Naturally, the phase-stability principle was an essential ingredient to its design, and Ed Lofgren converted the old 37-inch cyclotron on the Berkeley Campus to a one-quarter scale model of the Bevatron in record speed, less than one year!

I participated in various design discussions on the Bevatron, and Ed McMillan and I independently did the relativistic calculation to define the threshold energy for protons impacting a fixed target to produce anti-protons, if they existed. It was characteristic of the times that such a calculation, which is totally routine today, was at the time considered a substantial contribution.

The foregoing summary account illustrates a very productive period from roughly 1947 to 1950, when I was doing experiments and teaching without being significantly diverted by external problems, but this changed rapidly for two reasons. The first was the increasing concern about nuclear weapons induced by the worsening of the Korean War combined with the future of nuclear weapons during the ongoing Cold War, and the second was the increasing hysteria about Communist infiltration during the McCarthy era.

5
Military Work at Berkeley and the Loyalty Oath

Despite the unprecedented and highly productive postwar activity at UCRL in accelerator-based high-energy particle physics, E. O. Lawrence personally never quite demobilized after the end of World War II. He pursued the Calutron electromagnetic uranium enrichment program until the AEC terminated it in 1948. He maintained a skeptical aloofness from the efforts to achieve international control of atomic energy that were to be followed by worldwide prohibition of nuclear weapons. The initiatives of Robert Oppenheimer resulting in the Acheson–Lilienthal Report were converted into the ill-fated Baruch Plan, but Lawrence refused to serve on the advisory committee to Bernard Baruch, notwithstanding that the proposal would have perpetuated the U.S. monopoly on nuclear weapons and was, as expected, rejected by the Soviets.

Then the Soviets conducted their first nuclear test on August 29, 1949. Even before that event, Edward Teller, supported by Lawrence, exerted continuing pressure for an accelerated program to design and build a thermonuclear weapon (the "hydrogen bomb"). The advisability of proceeding on such a crash program was considered by the General Advisory Committee to the AEC in 1949, with those eight of the committee's members present (one member, Glenn Seaborg, was absent) all opposing the idea. The AEC had supported its General Advisory Committee advice with a bare three to two majority, but as a result of pressure both from inside and outside the administration, Truman reversed the AEC decision and ordered a crash program. All these events have been extensively documented,[1] and understandably had repercussions at the Berkeley Laboratory. They caused much internal discussion among those "in the know."

Luis Alvarez talked to me frequently about the necessity of rapidly proceeding with the H bomb, and maintained that if the Soviets attained such a device earlier than the United States, even for a short time interval, then America would be in serious danger. I remained unpersuaded by these arguments, recognizing that the United States at that time had already accumulated several tens of nuclear fission weapons, and that these would be an adequate deterrent.[2] Moreover, the enormous potential explosive power of hydrogen weapons generated many doubts in my mind on both their utility and morality. I once went home from working at UCRL and said to my wife, "Today is a red letter day; Luis Alvarez did not call me a traitor."

Nevertheless, I continued to engage in numerous deliberations addressing the potential contributions of UCRL to continued U.S. work on nuclear weapons. With the termination of the Calutron program, Lawrence promoted a UCRL-sponsored construction on a nearby site of a nuclear reactor designed for propulsion of naval vessels. This proposal was turned down by the AEC on the grounds that UCRL lacked experience in reactor design, and that if such a project should proceed, it should be executed under more experienced leadership.

Lawrence and Alvarez then turned to the potential UCRL contributions to generating large quantities of neutrons, which would be sufficient for either production of tritium or for breeding plutonium from the large stocks of depleted uranium accumulated from the separation plants in the East at Oak Ridge. Although at that time no workable design of a hydrogen bomb existed, tritium was believed to be an essential ingredient of such a device, and plutonium produced from existing reactors "burning" uranium might face a cutoff of the overseas supplies of uranium ore. Again, Lawrence and Alvarez proposed production reactors for this purpose, but after being discouraged by the AEC, they turned to accelerator production. Lawrence put Alvarez in charge of these programs, and Luis talked to me extensively about the technical options.

One candidate for accelerator production was the so-called sector-focusing cyclotron proposed in 1938 by L. H. Thomas.[3] Although that machine employed magnetic fields varying in azimuth, "strong focusing" had not been recognized at that time as a general principle. Partial utilization of what can be considered strong focusing had taken place previously; in addition to the sector-focused cyclotron, magnets of alternating gradients had been used before the war to focus the external beam of the Princeton cyclotron, and some applications in electron microscopes used what amounted to strong focusing.

Of course, mathematically, the properties of the harmonic oscillator equation with periodically varying restoring force had been well known for a long time as a result of the solutions of the so-called Floquet equation. That equation exhibited alternate bands of stable and unstable solutions. But using the sector-focusing cyclotron as a very high intensity neutron source appeared to offer undue risks: internal targeting for extremely high-powered beams appeared impractical, and beam extraction with the required high efficiency remained an unsolved problem.

Accordingly, Lawrence and Alvarez turned to the linear accelerator as the most promising approach, because the practicality of a proton accelerator had been demonstrated by the 32-MeV machine. Alvarez, in turn, worked with me on possible approaches. In the absence of recognition of the potential of "strong focusing," I designed a potential scale-up of the 200-MHz machine to lower frequencies using magnetic solenoids incorporated into the drift tubes as focusing elements. Because the focusing forces produced by magnetic solenoids are relatively weak,[4] this required a very large aperture in the drift tubes.

Alvarez and Lawrence envisioned a variety of stages to attain a neutron production plant producing perhaps a gram of neutrons per day! The first stage chosen was the MARK I stage, code-named the Materials Test Accelerator, or MTA. The design responsibility for the MTA was largely divided among Ed Lofgren designing a deuteron ion source, myself designing the accelerating cavity and analyzing the orbit dynamics, and Harold Brown (who later was to become secretary of defense) designing the targets. Herbert York had measured the expected neutron yields from deuteron beams of various energies. This team produced a design of promise sufficient for Lawrence to secure approval from the AEC for its construction. While the design work was carried out in a classified area in the Berkeley UCRL site, unbeknownst to me, Lawrence selected an abandoned naval air station near Livermore, California, for the construction site.[5]

The Livermore site was initially operated as an adjunct to UCRL; it only later became an independent laboratory as the second U.S. nuclear weapons national laboratory (LLNL).

Because of the large diameter of the beam as focused by the solenoids, the frequency chosen for the accelerating cavity was 12 MHz, which in turn gave the linear accelerator cavity a diameter near 60 feet. This relatively low frequency was also chosen because oscillators of adequate power were not at the time available at higher frequency. The length for this "pilot model" was chosen to be 87 feet, truly a giant challenge to vacuum engineering. Lawrence persuaded the management of the Standard Oil Company of California to create a subsidiary designated the "California Research and Development Corporation" (CR&D) to provide the engineering and construction of the MTA. In addition to designing the fundamentals of the RF accelerating system, I was enlisted to give lectures to the Standard Oil engineers on nuclear physics, and on the penetration of particles in matter. They were highly capable people in their fields, but had no previous contact with nuclear matters.

Technically, the machine turned out to be successful. It produced continuous proton beams of unprecedented power. A major difficulty was that the energy stored in the huge electromagnetic cavity was so large that any discharges proved to be highly destructive. If such discharges occurred, they would produce spectacular "stalagmites and stalactites" of copper; therefore the gradient had to be restricted to avoid such events. The gradient required for the structure to accelerate protons was 1.25 MV/meter, whereas it was 2.5 MV/meter for deuterons. Therefore, the use of the MTA had to be restricted to proton acceleration. The maximum beam continuous current obtained was between 0 and 100 milliamps (ma) of protons at 12 MeV, and at the peak current was 225 ma at a 20% duty factor. Because the range of protons in solid material at this energy is only a few thousandths of an inch, stopping a beam of such power was extremely difficult and required precessing the beam at high speeds, using a rotating magnetic field, similar to that in the stator of an alternating current electric motor.

The machine went into full operation after I left Berkeley in 1951. Several other, shorter models were built after I left, but then the entire program of accelerator "breeding" of nuclear materials was cancelled. The main reason for this program had been to protect against a possible shortfall of natural uranium; that mission disappeared once serious uranium exploration commenced in this country, uncovering very large resources in Colorado and elsewhere. Thus, the large follow-on production machines to the MTA were never built, and that machine was turned over to the AEC's Division of Research for work in basic nuclear physics. As a result, some useful experiments were done, but the extremely high operating cost of the MTA forced its closure in 1953.

I participated in some of this design work on the MTA with considerable misgivings in parallel with the much more pleasing and creative work on the machines on the Berkeley site and with my teaching on campus. By that time, I had been appointed an associate professor with tenure in the physics department. But then, on top of this major diversion of the work of UCRL to the Cold War, came the impact of the McCarthy era, which generated suspicion of Communist infiltration in government and other major institutions.

In 1949, the resident representative of the University of California to the state government in Sacramento reported to the president of the university and the Regents that the California legislature might pass drastic measures prohibiting the employment of faculty and staff members who had a Communist background. To preempt such moves, President Robert Gordon Sproul proposed that a loyalty oath be taken by all university employees affirming their lack of Communist contamination. This proposal, which the Regents endorsed, created enormous controversy on campus and, to a somewhat lesser extent, also on "the Hill" at UCRL.

Some of us at UCRL who held security clearances, and were therefore accustomed to the kind of irrationality and lack of privacy inherent in personnel security measures, reluctantly signed the oath, although we thoroughly disapproved. Others, and particularly those of European background, refused to sign, noting that Mussolini, for instance, had used a loyalty oath to shore up the Fascist regime in Italy.

When the loyalty oath controversy degenerated to the point that tenured faculty members and other staff who refused to sign were threatened with dismissal, I decided I could not tolerate the situation further. I informed Lawrence, Alvarez, and other colleagues of my intent to leave the Radiation Laboratory. Lawrence proposed that before making a final decision, I should talk to John Francis Neylan, the president of the Board of Regents of the university "to hear the other side." Lawrence, who maintained extensive contacts with the Regents, persuaded Neylan to receive us at his Atherton estate, and Lawrence drove me there while we discussed the situation in the car.

After I sat down with Neylan, he said to me, "Young man, what is bothering you?" And I replied that the rights of those whose conscience made signing the oath unacceptable should be respected. Neylan in turn embarked on a long talk which I found interesting and somewhat surprising. He said that he really didn't have trouble with Communists being part of the faculty or staff of the university, and that he in fact had some friends who were Communists. However, he felt that the faculty was "acting irresponsibly" by continuously changing position and proposing compromises. Indeed, the faculty, under the chairmanship of the prominent chemist Joel Hildebrand, had agonized during several extended meetings as to what to do in response to the proposed oath. Anyone familiar with faculty conduct should not have been surprised that inconsistent and variable positions emerged. At any rate, Neylan continued his monologue and I was unable to make any further remarks. Lawrence drove me back to Berkeley, and I said that I had not changed my mind.

The fact that I was planning to leave UC Berkeley became known, together with the plans of a significant number of other faculty members to depart. The university, including UCRL, lost many highly prominent members, including Gian-Carlo Wick, Jack Steinberger, and Geoff Chew. Bob Serber left some time thereafter for Columbia University; many young theorists departed also. The loyalty oath was invalidated by the courts some time later.

Before my departure, the security people at UCRL asked me for a final interview "to clear my record." They asked two questions. First, did I know that I had walked into a "subversive" bookstore while in Pasadena, and secondly, did I know Robert Andrews Millikan? I replied to the first question that I had visited many bookstores in Pasadena and had no idea which ones, if any, were subversive. I expressed surprise in response to the second question saying that I knew Robert Millikan very well because he had first invited me to become a graduate student at Caltech, and because he had been very kind to me on many occasions thereafter. I attended a one-semester course he gave on atomic physics. The security people asked whether I knew that Millikan was a member of something called the Society for American–Russian Friendship. I replied that I did not know this. This episode was highly amusing to me because Millikan, aside from his accomplishments as a physicist and Caltech president, was well known to be an extremely conservative, very religious family man. In addition, he was very solicitous of the welfare of the younger members of the institute, including myself. After the interview, I reported to my wife that there must be something terribly wrong with me if I didn't know anyone more subversive than Robert Millikan.

I received a number of offers from other universities, and also from industry. But then a delegation from Stanford University consisting of Leonard Schiff and Felix Bloch called on me in Berkeley and proposed that I accept a professorship in the Stanford physics department. They talked to me extensively about the plans to complete the one-GeV electron linear accelerator at Stanford and the opportunities this would offer. I knew very little about

Stanford at the time. I had heard, of course, about the famous work of Felix Bloch on nuclear magnetic moments, and had attended meetings of the American Physical Society at Stanford. I was also familiar with the collaboration between Alvarez and Bloch on the classical experiment measuring the magnetic moment of the neutron.

During a seminar at UC, Edward L. Ginzton, the director at Stanford's Microwave Laboratory, had explained the electron's motion in a traveling wave linear accelerator using a "surfboard analogy": the electron, likened to a surfboard rider, glides always "downhill" on a wavefront which, however, travels along at the same speed with the electron. Bob Serber, who had worked extensively on the theory of particle motion in linear accelerators, reacted: "I always did want to know how a surfboard works."

I also recall being present at a meeting between William W. Hansen, the great Stanford physicist, with Luis Alvarez sitting on a bench outside at UCRL. Hansen had invented the disk-loaded waveguide as a means to accelerate electrons. He had succeeded in doing so in 1947 using his "MARK I" 3.6 meter machine which attained 4.5 MeV. Luis Alvarez insisted that a large electron linear accelerator could not possibly work because electrons, once approaching the speed of light, could not be "phase stable;" that is, they could not adjust their velocity to match the phase of the accelerating field. Hansen insisted that phasing could be accomplished by direct control, but Luis felt this was impossible.

After mulling over the situation in Berkeley I decided to accept the Stanford offer. Somehow, moving across the Bay with my expanding family was highly attractive, and was probably a psychological indicator that I really wanted to remain at a short distance from the work of UCRL. Luis Alvarez tried very hard to dissuade me from going, and stated flatly that I would never be able to do any physics again. But despite all these controversies, Luis and I maintained our friendship, and I interacted with him many times in the years to come. I also maintained my friendship with Ed McMillan throughout this period and afterwards.

6
Beginnings at Stanford

We crossed the San Francisco Bay from Berkeley to the Peninsula in the beginning of July 1951 with what was by then a family of six. We also arranged for a moving company to move our stuff. On arrival at our new home, the movers refused to unload our household goods until they received payment. My wife was holding our latest addition to the family, and said she could not reach her pocketbook until the movers had unloaded the crib so that she would have a place to put the baby down. This resulted in an impasse that was resolved by Adele saying, "OK, you hold the baby, and I put up the crib," to which the mover replied by assembling the crib.

We had selected a home about six miles south of the Stanford campus with the kind help of Ed Ginzton, then director of the Microwave Laboratory, and V. L. VanderHoof, who had by that time returned to his professorship at Stanford. This was a large, somewhat dilapidated house built in 1907, one year after The Great Earthquake. Although it was built of excellent materials, it needed and received a great deal of work; it has served us well ever since that time. In particular, it permitted us to accommodate our growing family and to conduct seminars and other meetings at home. We specifically made this choice to live off-campus; we felt that it would be preferable for our kids to be educated outside an all-academic setting.

My work at Stanford encompassed research, lab direction, teaching, and public service. These activities largely overlapped in time, but in this book, I account for each of them separately, in the interest of continuity. I recount my research and lab direction activities here, and my teaching and public service experiences in later chapters.

I joined the Stanford faculty at the time of a major expansion of the university. In fact, that expansion was unprecedented in that it moved Stanford into the ranks of the foremost universities in an amazingly short period. Much of this has been extensively documented. For an account of that growth and the contributions of Provost Fred Terman toward that goal, the reader is referred to Gillmor's book describing Terman's role in the service of the university.[1]

I worked at both the Department of Physics of the university, and what was at that time the Microwave Laboratory. I arrived at Stanford during the summer break, and therefore could get started addressing the primary problems at the Microwave Laboratory's high-energy physics accelerator programs. Teaching at the Physics Department did not start until later in the fall.

The Microwave Laboratory housed two accelerators, the MARK II and the MARK III, the latter being the machine proposed by W. W. Hansen in March of 1948 to reach an electron energy of 1 GeV. Hansen can justly be credited with being the initiator of microwave and electron accelerator activities at Stanford. He was an extraordinarily gifted physicist, an excellent theoretician interested in basic processes, and at the same time a "hands-on" experimentalist. I remember him saying, "Don't give me another engineer; give me another machinist." I knew him through casual contacts only; sadly, he died in 1949 from chronic lung disease one year before I joined the Stanford faculty. Hansen had been a good friend of my mentor and father-in-law, Jesse W. M. DuMond. He was truly the "uncommon man" to quote an award that he received. The chain of progression of the electron linear accelerators started with Hansen's prewar invention of the electromagnetic cavity or rhumbatron, generated by his realization that the separate elements of capacitor and inductor in a conventional resonant circuit could be combined into a single resonant entity.

Although that invention was originally motivated by the desire to accelerate electrons to high energy, it was subsequently used for the generation of microwave power during Hansen's collaboration with the Varian brothers, Russell and Sigurd. The Varians were research associates at Stanford working together in a small room in the physics department; their collaboration with Hansen resulted in the invention of the klystron tube. Hansen and the Varian brothers gradually accumulated a group of young collaborators and subsequently the klystron played a substantial role during World War II in radar devices and other microwave applications. This entire Stanford group migrated during World War II to the Sperry Gyroscope Company, a major defense contractor in the East. After the war, Sperry did not wish to create a branch on the West Coast, so the Stanford contingent returned, some of them rejoined academic life and others started their own company in 1948, Varian Associates. But Sperry continued to support the activities of the physics department in exchange for shared patent rights.

All this has been a classic story of how ideas starting from fundamental concepts can evolve into a major and successful undertaking. The story is well documented elsewhere; in particular, in a personal account by E. L. Ginzton.[2]

During the war, klystrons served principally as local oscillators in heterodyne circuits at power levels in the milliwatt range, but some applications had extended klystron power close to 20 kilowatts. However, to power the large electron linear accelerator proposed by Hansen in 1948, tubes of peak power in the tens of megawatt range were required. Thus, the successful development of the MARK III accelerator required two new advances. The first was the ability to greatly extend the manufacture of the linear accelerator. This could be accomplished by improving on Hansen's experience with the MARK I.

A small machine built by Hansen in 1947, the MARK I had led to his famous, shortest report of government-sponsored research ever written, reading in toto, "We have accelerated electrons." At the same time, a group led by Professors Edward Ginzton, Marvin Chodorow, and others tackled the second necessary advance: extending klystron power by a factor of several thousand. These developments, which required and achieved enormous extension over past experience, had to be assembled into a working laboratory.

A detailed account of the status of the MARK III accelerator and its ancillary facilities was published[3] in 1955. That paper describes the status when the maximum energy attained by the accelerator was 630 MeV, and gives a considerably fuller account of the history preceding the project and supplies full detail of the design and construction of the accelerator, including a discussion of principles of operation, tolerances, and so forth.

I therefore do not recount these dramatic developments here in any detail, but describe the "challenging" situation I met when I arrived at the laboratory.

There were many problems. The linear accelerator would not tolerate as high a gradient as was anticipated, and therefore it had been extended for the full length of the building in order to attain the 1-GeV energy goal. No room for any experimentation had been left at the end of the accelerator. Part of the reason for this limitation was that the traveling wave accelerator sections were built by the "expand and shrink" method. The disks were stacked on a mandrel with highly precise spacers. They were then immersed in liquid nitrogen, and inserted in a precisely machined copper tube that was warmed in a steam jacket. When this whole assembly reached room temperature, the disk would expand and produce a shrink fit with the copper tube. However, over the operating life of the machine, problems were encountered with cold flow, which would relax the contact between the disks and the cylinder, leading to arcing at that point of contact.

There were also problems with the manufacture of the klystron tubes and the reliability of the modulators that were triggered by a spark gap switch. All of this had been addressed during the development cycle of the machine, but at the time of my arrival, Ginzton's attention was being divided between managing the construction and commissioning of the accelerator and producing klystrons for a large variety of other applications. Therefore, the day-to-day management of the accelerator construction and commissioning was the responsibility of R. L. Kyhl, an extremely able physicist and engineer, but who unfortunately already suffered from some health problems. In addition to Kyhl, Ginzton had assembled a very able team, so therefore solid progress was achieved, and the energy attained by the machine steadily advanced.

Richard Neal, after serving in the Navy, wrote his Stanford Ph.D. thesis[4] on the construction of the MARK III machine. Richard F. Post, as part of his Ph.D. thesis, supervised the construction and operation of the short (4-meter) MARK II accelerator which served both as a testbed for the larger machine and as a very productive accelerator for photo-nuclear research, largely directed by W. C. Barber.

A thesis by En Lung Chu described the theory of linear accelerators in extreme mathematical detail.[5] I found that his thesis, although extremely scholarly and elaborate, was difficult for the practioners to use. I therefore published several notes on simpler versions of linac theory.

Governmental support for the accelerator work at Stanford was provided through the Office of Naval Research (ONR). Its head, Emanuel (Mannie) Piore was a very farsighted individual who recognized that the flood of students returning to the universities after World War II meant that universities would have few resources of their own to dedicate to research. In addition, the wartime contributions of physicists convinced ONR and the AEC that physicists, if adequately supported, could execute complex undertakings. Whereas the Atomic Energy Commission largely concentrated its support on its own laboratories, ONR supported university research on a broad front. Piore had been particularly impressed by Bill Hansen's accomplishments. Even after Hansen's untimely death, Piore provided continuing support for the accelerators and microwave efforts at Stanford.

The Microwave Laboratory carried out some classified work, mainly consisting of designing and supplying klystrons for radar applications. This led to some absurd situations. For instance, a series of one-megawatt tubes was produced and painted in two different colors, one of them being classified, the other being unclassified. The situation was further complicated by the cancellation of Ed Ginzton's security clearance. The matter was resolved, but not without a hearing as described by Ginzton in his memoirs.[2] I intervened in the matter with the Chief of the Office of Naval Research, pointing out the absurdity of the situation.

Professor Robert Hofstadter had been recruited from Princeton University to join Stanford in 1950, and with the assistance of able collaborators, was intensively pursuing his program of studying the scattering of electrons from diverse nuclei. Inasmuch as commissioning of the MARK III machine was proceeding slowly, and there was no room for experiments at its end at any rate, he first built a small spectrometer to be located at the halfway point of the accelerator. His spectrometer was a scaled-up version of an instrument used for beta-ray spectroscopy at Caltech, and it exhibited point-to-point focusing between target and detector. Hofstadter generally used thallium-activated sodium iodide crystals as detectors which he had discovered to be very effective γ-ray and electron detectors during his earlier work.

When I arrived at Stanford, there was little communication between Hofstadter's group and the accelerator builders, quite apart from the various physical and technical problems mentioned above. One immediate technical issue derived from the fact that the energy and energy spectrum of the machine strongly depended on its operating conditions, such as phasing and the capture of stray electrons picked up along its path. I concluded that any practical experimentation would require a magnetic filter system that only permitted passage of electron beams of preset energy and energy width. I therefore constructed what was called a magnetic achromatic translation system[6] which was inserted at the halfway point in order to supply Hofstadter's experiments. In addition, it was clear early on in my tenure that a building extension had to be designed and constructed in order to permit experimentation at the peak energy of the MARK III accelerator.

Extending the building became my responsibility in interaction with the university's architectural department. It needed to be a very rapid undertaking; but because of the specialized application, the university architects were surprised by the need for a great deal of interaction with me in order to proceed with contracting the job. Financial support was arranged largely by Ed Ginzton tapping some of the royalty income accumulated from the klystron patents.

One feature of the building extension, also called the "end station," was its beam switchyard. It incorporated two achromatic translation systems, thus producing two beams for use in the end station. The beam switchyard was designed in cooperation with Jack McIntyre, one of Bob Hofstadter's team members.[7] A large mound of earth generated by the excavation for the end station and beam switchyard also served as a beam stop behind the building. Providing two beams into the end station was deliberate: one of the beams was intended to supply Bob Hofstadter's experiments on electron scattering, and he proceeded to design a much larger spectrometer than the one he had built at the halfway point of the accelerator. That spectrometer was a further scaleup of the semicircular design originally developed at Caltech. The second beam was intended to supply a large variety of experiments. Figure 6.1 illustrates the arrangement.

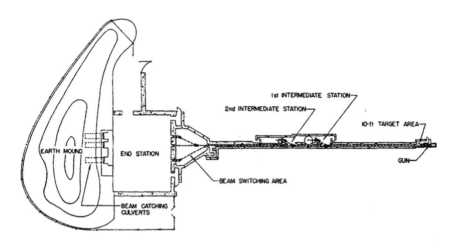

FIGURE 6.1. Layout of the MARK III electron linear accelerator and its beam switchyard and end-station at the Stanford High Energy Physics Laboratory. (From M. Chodorow et al.[8])

Because of Ed Ginzton's rather free-wheeling business practices, Leonard Schiff recruited Frederick V. L. Pindar to run the business affairs of the Microwave Laboratory, an excellent choice because Pindar kept us all honest while accommodating our evolving technical decisions. Pindar had a very large responsibility. The work of the Microwave Laboratory was funded from diverse sources and costs had to be controlled. Property had to be tracked and purchases had to be executed under the many programs facing tight deadlines.

One interesting episode is worth noting. The Navy supplied lead ingots weighing about 100 pounds each which formerly served as ship ballasts. These were cast by the laboratory into lead bricks that could be stacked to serve as radiation shields. The ingots were stored in a fenced enclosure outside the laboratory. We noticed that the size of this lead stockpile kept shrinking, and our night guard was "staked out" in a car to observe this phenomenon. The pile continued to diminish and apparently the guard was asleep during his stakeout. Then Pindar himself, with a large pot of coffee, carried out the watch for many nights. He observed a truck with a double bottom approaching, and a large man dug a trench under the fence, removed some lead ingots, and backfilled the trench. The thief then ran, carrying a one-hundred pound ingot under each arm (!) to his truck. Pindar called the police, and the thief was arrested. At the arraignment hearing, the judge asked for $10,000 bail, which was paid in cash. The judge said he would prefer a check, but the man said that he did not trust banks. The culprit was tried, convicted, and sentenced to several years in prison, and he swore at his conviction that upon his release he would kill Fred Pindar. Happily, this threat was not executed. The stolen lead was located in the perp's junkyard and was returned to the laboratory.

In view of all these complications, Ginzton and I decided to declare a friendly divorce. The Microwave Laboratory was divided into the High Energy Physics Laboratory (HEPL), of which I became director, and the Microwave Laboratory, directed by Ginzton. Both laboratories together were named the W. W. Hansen Laboratories, and Fred Pindar became the chief administrative officer for the combined laboratory complex. Because of the magnitude of the job, he was joined by Marshall O'Neill in 1952 and by E. B. Rickansrud several years later, both of whom continued to play major administrative roles in later university endeavors.

The physics department at Stanford had an excellent record of scientific achievements, yet it was a very conservative institution. This was due in part to the influence of Felix Bloch, who attempted to mold the department and its professors in the European tradition. The department exercised what I called a "Noah's Ark policy," that is to say, "two of a kind." The department generally had two professors working in each of the major disciplines of physics. There was some tension between the physics department and the university administration on the question of salary splitting with the government. The physics department insisted that the salaries of all academic professors during the regular term be paid fully from "hard" university funds, irrespective of whether they were engaged in government-supported research; only summer salaries, if professors chose to continue their work during the summer recess, could be reimbursed from "soft" government funds. In contrast, many other university departments, in particular those in engineering, were quite willing to accept salary splitting during the regular term, thereby gaining greater flexibility in funds for expanding their faculty.

This problem was to some extent a consequence of the checkered history of the Department of Physics. William Webster, Leonard Schiff's immediate predecessor as chair of the department, had encountered numerous conflicts between "pure" and "applied" physics when the Sperry Company was providing funds in support of physics research, while at the same time the department had been struggling to maintain its "purity." These problems eve .tually resulted in separating the applied physics members of the physics department into a division of applied physics, which later became a full-fledged Department of Applied Physics. As a result of this and other arrangements, "Physics at Stanford" was divided among the physics department, applied physics department, and also electrical engineering. The latter department included excellent work in radio astronomy, microwave tubes, and some solid-state physics. The Microwave Laboratory and then the High Energy Physics Laboratory were formally independent laboratories, but the understanding was that the scientific policies of HEPL were to be controlled by the physics department.

Notwithstanding all these complications, the physics department offered a highly collegial and friendly atmosphere. Part of this was fostered by the tradition that teaching responsibilities were to be allocated equitably among the professoriate irrespective of the research or administrative responsibilities of individual professors. In addition, according to department policy, the teaching of the introductory courses was to be shared by all senior professors, again irrespective of their research activities or other responsibilities. Thus beginning undergraduates generally enjoyed extensive contacts with prominent professors active in research, a rather unusual pattern in a major university.

At the same time, appointments of new professors rested almost entirely on demonstrated research accomplishments and promise, not on teaching performance. As a result, some of the lectures given to undergraduates left much to be desired. The teaching versus research issue came to a head when the physics department appointment committee (consisting of Professors Schiff and Bloch) rejected the candidacy of a candidate with highly luminous teaching credentials but a lackluster record in research. This rejection, endorsed by a majority vote of the faculty, resulted in an outcry by one of the department's most prominent teachers, Professor Paul Kirkpatrick, a close friend of mine and Jesse DuMond's, as follows. "The mail is heavy with outgoing manuscripts and incoming honors, but who worries about teaching?" But the decision stood.

7
Research and Teaching Before SLAC

After completion of the end station and gradual upgrading of the MARK III accelerator's energy and general performance, it was possible to engage in a full-fledged research program. While this proceeded, I was also becoming more engaged in military and arms control issues and my teaching began in earnest following the policy of the physics department that teaching responsibility should be broadly shared among the faculty.

In retrospect, I find it difficult to recall how all these responsibilities became compatible with life at home. But they did. In our house, the upper story was dedicated to the kids, and my wife and I shared a bedroom downstairs because there were many night shifts at the laboratory. This arrangement permitted me daytime sleeping. In addition, we managed many great family excursions to the mountains and also some trips East to introduce my children to my parents.

At the laboratory, we were able to institute regular scheduled operations. Several experienced people came with me from Berkeley. Robert Mozley joined me after his rather exciting experiences acquiring a Ph.D., and he helped a great deal in establishing reliable operations. Also, Carl Olsen, an engineer experienced in detector electronics came along, and so did Phyllis Hanson, an experienced technician. She had been married to a perennial graduate student at Stanford while she was at Berkeley, and she was highly pleased to avoid commuting by joining me at the HEPL enterprise. She became our lead operator, although difficulty arose in that I had designed such a narrow shielded entryway into the beam switchyard that she was incapable of entering when she was expecting her child.

HEPL was not operated as a facility for outside users. Rather, Hofstadter and I, both professors in the physics department, each used one of the two beams emerging from the switchyard. Bob Hofstadter rapidly proceeded with his comprehensive electron scattering program. For this program, he collected a group of capable postdoctoral associates. Hofstadter organized a "Conference on Nuclear Sizes and Density Distribution" during December of 1957 which attracted an international audience. As is well known, Hofstadter received the Nobel Prize in 1961 in recognition of his electron scattering work.

I pursued a less specialized and generally more exploratory program of high-energy physics in the second beam. I became convinced that high-energy, high-intensity linear electron accelerators could support a very broad program of work of which elastic electron scattering was an important, but not the only, component. In the past, research at other electron machines, including the early Betatrons, had been relatively limited. This was also true to some extent in respect to the work on Ed McMillan's 300-MeV synchrotron that I have previously discussed. The work in the second beam was mainly carried out with a large group of physics department graduate students, 13 if my present count is correct.

The first experiment was carried out with a graduate student[1] before HEPL was formally established, and before the end station became available. This was a determination of the integrated cross-section for the production of N^{12} by the $(\gamma, 2n)$ reaction in Nitrogen 14. This reaction was eminently suitable for a "quickie" experiment because the short half-life of this isotope, 12.5 msec, permitted detection between pulses of the accelerator with an improvised beta ray spectrometer. The resulting experiment may be of some current practical interest as a means of detecting smuggled explosives. This experiment gave me my first experience with the priorities then prevailing among my professorial colleagues. We had scheduled a run on the experiment, and the graduate student, D. Reagan, did not show up. I found that he was at the home of a Stanford professor and I called there. He came to the phone, and I reminded him of the run. A voice in the background, which I recognized as that of Professor Paul Kirkpatrick, yelled, "Tell Panofsky we're doing something important!" They were conducting a rehearsal for the physics department Christmas party of a bottle orchestra, which used glasses and bottles filled with water to be tuned to make tones of various pitches.

A broad area of attack was the study of production of pions directly by electrons to complement the production processes by γ-ray photons that had been undertaken at Berkeley. The short duty cycle of the machine did not encourage observation of coincidences between the detected pions and the inelastically scattered electrons, so the pions only were detected in most of the experiments. Not surprisingly, most of these results were equivalent to those obtained in photoproduction because the predominance of the cross-section was generated by electrons scattered at very small angles. Therefore some of this work simply confirmed the concept of scattering of electrons constituting an "effective photon" as had been earlier promulgated by the Weiszacker–Williams approximation. Moreover, the photo-pion experiments filled in numerous gaps in the knowledge of these processes and demonstrated the power of electron accelerators. The first paper,[2] written in collaboration with two graduate students, gave a quantitative measure of the number of equivalent photons responsible for pion production. The quantitative value of the equivalent photon number depends on whether the photon absorption is an electric or magnetic dipole transition.

A further paper[3] measured the angular distribution of pions produced in hydrogen by electrons (or photons) where there had been conflicts in predictions among theories. We designed, principally in collaboration with Karl L. Brown, a "double focusing zero dispersion spectrometer" for measuring the charged pions produced.[4] This name seems to be a contradiction in terms: how can a spectrometer have zero dispersion? The instrument as built provided dispersion between two magnets, and therefore, slits introduced into the gap between the two magnets could define the energy band. The orbits were then refocused, making it possible to use very small detectors, thus reducing background.

We carried out one challenging experiment to examine electron production of pions by observing the inelastically scattered electrons. The kinematics of the scattered electron were chosen so that the pion was produced at the so-called 3–3 resonance, that is, the first excited state of the proton of isotopic spin 3/2 and spin 3/2. That state had been extensively explored previously in the photoproduction experiments in Berkeley and the pion scattering experiments in Chicago. The idea of this experiment[5] was to examine the cross-section of producing that resonance as a function of the four-momentum transfer to the proton by scattered electrons. Thus in effect, this experiment compared the size of the 3–3 resonant state with the size of the ground state of the proton as had been measured by Hofstadter and collaborators. The result is shown in Figure 7.1. This figure

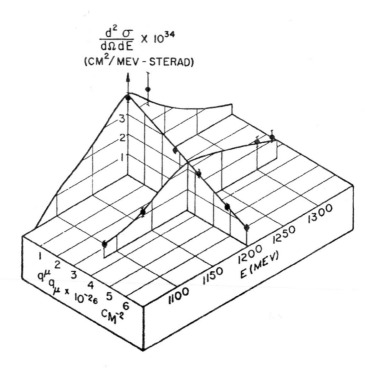

FIGURE 7.1. The differential inelastic cross-section for electron scattering from hydrogen leading to the 3–3 resonance on the proton, as a function of the four-momentum transfer and the electron energy. (From W. K. H. Panofsky and E. A. Allton.[6])

illustrates the shape of the resonance both as a function of energy, while at the same time demonstrating the descent in cross-section as a function of four-momentum transfer. Possibly even more significantly, this was the first experiment on inelastic scattering on the proton, a process that was explored much more extensively later at SLAC.

Another experiment was designed to observe the electromagnetic production of muon pairs. Although production of electron–positron pairs by photons was of course well known, it had not been established at the time whether the same process also occurred for the heavier leptons: muons. The experiment, carried out jointly with George E. Masek, then a graduate student, demonstrated[7] a production peak of muons in the forward direction, much larger than could be accounted for by π–μ decay. Obviously this was, in essence, a confirmatory experiment: it would have been surprising if this peak originating from forward muon pair production had been absent.

A separate experiment—which turned out to be quite difficult—was to measure the absolute value of the electron–proton scattering cross-section at a very small angle, and thereby measure the radiative correction to the electron–proton scattering process.[8] That experiment showed that experiment and theory agreed within 2% where the theory included a radiative correction of around 4%, not a very spectacular result, but again a confirmatory experiment. The technique used was similar to the one used in the earlier proton–proton experiments at the proton 32-MeV linac which had observed the proton recoils by using a camera on which photographic plates had been mounted.

Much of the foregoing work was of considerable interest to the theoretical faculty members in the Department of Physics. Particularly, Professors Leonard Schiff, Don Yennie, and Sid Drell were of great help in working with me on the theoretical predictions.

I ventured into a somewhat amateurish theoretical calculation of the expected "radiation length" in molecular hydrogen as compared to the radiation length in monatomic hydrogen. In the past, calculations of the radiation length in liquid hydrogen had been based on monatomic hydrogen while the hydrogen molecule is H_2. That difference turned out to be relatively small, but was confirmed quantitatively in an experiment,[9] albeit at marginal statistical precision.

In addition to these undertakings in particle physics, I became interested in some instrumental developments. One was using microwave cavities to serve as deflectors of particle beams in order to provide time-of-flight measurements. Naively, I thought that cavities excited in the TE mode would provide such a tool. However, analysis[10] gave the surprising result that electric and magnetic deflections in such a cavity would cancel identically; in contrast, TM modes provided a deflection independent of the particle velocity. This theorem later found some wider applications.

The transverse deflections in such a cavity were put to the test by P. R. Phillips[11] who built a time-of-flight detector. The cavity was introduced into the mid-plane of the double-focusing spectrometer described above and behaved as theoretically predicted. This was a complex experiment involving extensive microwave "plumbing" to achieve synchronism with the beam.

A further instrumental development was the wide-aperture rectangular magnetic quadrupole.[12] Conventional quadrupoles are constructed with hyperbolic iron pole pieces that constrict the aperture of the device. The rectangular quadrupole, as developed, permits an exact mathematical solution for the field, essentially independent of the saturation properties of iron. However, it does so at the expense of doubling the power consumption relative to a quadrupole producing a comparable magnetic gradient.

A further experiment was carried out to confirm the so-called Überall Effect which enhances the bremsstrahlung produced in crystal targets. In the usual bremsstrahlung process, the momentum transfer to the nuclei of the radiator material is quite small, and is nonzero only because of the finite mass m of the electron; in fact, the momentum transfer varies as m^2/E where E is the incident electron energy. If the wavelength corresponding to the momentum transfer to the nuclei constituting the crystal corresponds to the Bragg condition for x-ray reflection, then the radiative process is enhanced. More precisely, if the momentum transfer equals a reciprocal lattice vector of the crystal, then enhancement is obtained. This prediction had previously not been tested experimentally. We constructed a goniometer on which a single crystal of silicon was mounted.[13] The x-ray intensity was studied as a function of angular orientation of the goniometer relative to the incident electron beam direction. Enhancement was indeed observed, but it proved impossible to demonstrate quantitative agreement between the observed enhancement and theory.

Beyond this large variety of experimentation carried out with physics department graduate students, let me turn to the searches for deviations from the theory of quantum electrodynamics (QED). These experiments were initiated at HEPL, largely by future Nobel Laureate Burton Richter, who had joined the laboratory as a research associate in 1956. The relevant Feynman diagrams for processes studying the validity of QED can include either virtual photons or virtual electrons. If the momentum transfer implied by these virtual lines were large enough, conjectures were promulgated that deviations from QED might occur and be observable; however, the energies accessible to the MARK III were sufficiently low that it would indeed have been surprising had positive effects been found. A process that implies a large virtual electron propagator is wide-angle electron–positron pair production by a photon. Burton Richter carried out such an experiment at HEPL, obtaining a null-result. The maximum momentum transfer reached in this experiment was 80 MeV/c, leading to the most sensitive QED test at the time. Later experiments along these lines at MIT caused considerable excitement when a false deviation was observed, but a still later experiment by Sam Ting again produced a null-result.

To examine a possible deviation from a QED process involving a diagram with a large momentum transfer photon propagator, an experiment on electron–electron scattering was proposed. This cannot be done by electrons striking a stationary target because the momentum transfer would be much too small; rather, a colliding beam arrangement is required. The initiative for such an experiment at HEPL, using the MARK III accelerator, originated from Gerard K. O'Neill of Princeton University, and a detailed proposal was developed by O'Neill, Burton Richter, and W. C. Barber; the latter had largely been working at the smaller MARK II accelerator at HEPL. That proposal initially generated substantial opposition from the physics department because HEPL was not chartered as a facility for outside users and as it was originally written, no Stanford faculty member was participating in it.

To circumvent this objection, I joined the proposed project [14] although I had relatively little to do with its development. The experiment required explicit approval from ONR because it would add substantially to the cost of operating HEPL, and this approval was granted. Construction of this colliding beam facility was a very extensive enterprise introducing much innovation into the work of HEPL. The experimental layout is shown in Figure 7.2. An ultrahigh vacuum system was built and tested at Princeton, and the magnet was constructed at HEPL. My only marginal technical involvement was in some of the design details of the double ring magnet. The detector was also a novel development and surrounded the electron–electron collision point with a large solid angle.

This pioneering experiment was completed slowly and did not obtain results until SLAC was already under construction. The final result was negative, but set a new limit on the validity of QED. In the process, O'Neill and

FIGURE 7.2. The electron–electron storage ring collider at the High Energy Physics Laboratory. (Credit: Stanford University photo.)

Richter realized that a storage ring providing for electron–positron collisions, rather than the electron–electron collisions they were now using would provide a much more powerful tool for a much wider group of experiments. But realizing such a program had to be postponed until SLAC became operational.

Another experiment worth citing here is the one by Kenneth Crowe and two collaborators to measure the beta decay spectrum resulting from stopped positive muons. That spectrum had been parameterized by L. Michel, and past measurements of the so-called Michel Parameter ρ had spanned a wide range between .23 and 0.6. Crowe and company reported $0.5 \pm .1$[15] for this parameter. Improved data reported at the 1956 Rochester Conference gave $\rho = 0.6 \pm 0.05$.[16] Today the value predicted by the Standard Model is $\rho = .75$, and recent measurements accurate to nearly one part in a thousand agree with that value. The experiment is of interest because for the first time an electron accelerator was used as a source of secondary particles for use in an experiment examining the properties of the particle generated.

Richard E. Taylor, the future recipient of the Nobel Prize, carried out an experiment as a graduate student under the supervision of Robert Mozley. Because of the formal rules of the physics department, Mozley was not authorized to actually accept Taylor's thesis, so I had to co-sign the work although I was hardly familiar with its content.

The totality of all these successful experiments confirmed that the electron accelerator had a potential to serve a very large variety of missions in experimental high-energy physics, thus complementing what was believed at the time to be the exclusive province of hadron accelerators. It was that experience which served us well in extending the future of electron accelerators at SLAC. I note that even though the work of HEPL was supported by ONR, the Atomic Energy Commission staff was well aware of its productivity. I was given the Atomic Energy Commission's Ernest Orlando Lawrence Award (Prize) in 1961 in recognition of my work at both Berkeley and HEPL.

I spent the summer of 1957 as a summer visitor to Brookhaven National Laboratory. Our family was quartered in the visitor's barracks on-site. The long-lived neutral K-particle (K_2) had been discovered by pictorial methods and I collaborated with Val Fitch to build a well-collimated kaon beam to permit electronic detection. This method worked and permitted the measurement of the absorption cross-section of the K_2; the answer turned out to be compatible with conventional strong interaction cross-sections.[17] Note that at the time, the short-lived and long-lived kaons were believed to be distinct particles of different parities, but accidentally of nearly identical mass.

During the summer at Brookhaven, we enjoyed the company of Richard Feynman in our barracks. He was in his compulsive joking mood and our children fell all over him. His shirt got torn up in the process and was patched with scotch tape. Dick reacted by saying, "Who tears my shirt tears trash."[18] It was an exciting evening complete with Long Island ducks.

Back at Stanford, in accordance with physics department policies, I engaged both in lower division and upper division teaching. For one quarter each year, I taught one course in the series in elementary physics for prospective engineers or scientists, including lectures on heat and light and on electricity and magnetism. The series had to be taught in threefold repetition because the lecture hall could hold only roughly one-third of the registered students. This was quite a chore because a large number of lecture demonstrations were involved. I recall that my wife helped me to write things on the blackboard in more legible script than mine, and also checked the line of sight from various vantage points in the lecture room to make sure that lecture demonstrations were visible. We designed some very complex demonstrations with excellent supporting technicians.

I remember once that one student habitually kept coming late by five to ten minutes to the great annoyance of the class. I had a demonstration where a light bulb was placed at a conjugate focus of a convex mirror, generating an image located above the real bulb, which was hidden. When the student again entered late, I asked him, "Could you please unscrew this light bulb?" He fumbled at the image without finding a light bulb, and then said to the great joy of the audience, "Well, I'll be damned." He never was late again.

The consequence of these lectures was that I was exposed to an enormous number of students and vice versa. A result is that whenever I travel these days, chances are very good that somebody will approach me at an airport saying, "I took freshman physics from you." And my usual answer is, "You poor guy."

I also taught upper division courses in classical electricity and magnetism —similar to the ones taught in Berkeley—and in classical radiation theory and an upper division laboratory. The American Association of Physics Teachers recognized this teaching effort by inviting me to give the annual Richtmyer Lecture in 1963.

All this seems to look like a heavy load, but it should be recalled that being director of HEPL was a light burden. Fred Pindar handled all budgetary matters and other administrative chores extremely capably so that the laboratory directors of the Microwave Lab and HEPL were pretty well isolated from these duties. In addition, Professor Hofstadter and I ran the programs in the two beams of the end station relatively independently, so there were not too many programmatic decisions to make. It was a very different world from that faced by laboratory directors today.

The profusion of results from HEPL, together with the past experiments at Berkeley, subjected me to many calls for presentations at international conferences. This was the time of the beginning of the annual so-called Rochester Conference on High-Energy Physics that had been started by Robert E. Marshak before I moved to Stanford. After that, international conferences in high-energy physics became institutionalized and I gave numerous talks both on the results of individual experiments and on surveys of the work at Stanford. But all these diversions were dwarfed over time by my becoming drafted to participate in numerous advisory bodies, both military and nonmilitary.

8
Science Advising and Arms Control: The Beginnings

The activities at Stanford University outlined in the previous two chapters were paralleled by my increasing involvement with various science advisory bodies and also in the international control of weapons.

I kept contact in Washington with both Atomic Energy Commission (AEC, the predecessor to the present-day Department of Energy) and ONR program officers for high-energy physics, although the work of the Microwave Laboratory described in the previous chapters was supported only by ONR. These officers were highly disturbed by the controversy concerning the crash program to proceed with H-bomb development and the subsequent—or possibly consequent—hearings conducted by the AEC that removed the security clearance of Robert Oppenheimer. That decision had a deep negative influence on the morale of some of the high-energy physics staff members in Washington. I then studied these events in some detail, and became profoundly troubled as well.

I have noted in Chapter 5 my reaction to the views of Luis Alvarez on the advisability of proceeding with a crash program on the hydrogen bomb: I was supportive of the "go slow" advice that the General Advisory Committee chaired by Oppenheimer unanimously gave to the AEC in 1949. In the later hearings, Oppenheimer was investigated not only for events relating to his past association with Communists but also for his expressed opinions, including those he held on the H-bomb program. Before Oppenheimer was appointed director of Los Alamos, his background had been well known by both the general government security establishment and by General Leslie Groves, the leader of the Manhattan Project. Oppenheimer's opinions were obviously not shared by all, but such divergences in view should in no way have been a factor in evaluating the extension of Oppenheimer's clearance. I therefore considered the verdict reached by the Oppenheimer hearings to be a travesty. As is well known, the presidential award of the Fermi Prize to Oppenheimer in 1963, initiated by President Kennedy before his assassination and subsequently bestowed by President Johnson, was an attempt to remedy that travesty.

Sometime before the AEC hearings on his security clearance were held, J. Robert Oppenheimer was asked in testimony before a congressional committee to explain how a nuclear weapon hidden in a shipping crate and smuggled into the United States might be detected, a question of very much interest today. His terse reply was, "With a screwdriver" (implying that the only way to detect a weapon in a crate was to open the crate and look). Subsequently, the AEC asked Bob Hofstadter and me to investigate how nuclear physics might be helpful in replacing Oppenheimer's screwdriver. Specifically, the question was: how would you detect one cubic inch of highly enriched uranium or plutonium if it were smuggled into the United States? Bob and I examined the passive radiations emitted by the materials in question and also made calculations of the results of irradiation of the containers, followed by detection of secondary emitted particles that might identify the contents.

This work resulted in the "Screwdriver Report," which was highly classified at the time, and perhaps still is. Interestingly enough, the broad conclusions of that report remain valid today. Indeed, you can detect passive radiations from plutonium, and to a lesser extent, from highly enriched uranium. Moreover, you can increase the sensitivity of such detection by various active methods, and we tabulated the combinations of doing so with various particles in and particles out. However, all these methods require close proximity to the object under examination, and the radiations, in particular those from uranium, can be reduced to escape detection by the use of shields of various kinds (some of which may be heavy). Since that time, these basic conclusions remain valid, although of course, the quality of both the available detectors and the data-processing power associated with analyzing the detected particles has enormously improved.

Our Screwdriver Report was followed by an attempt by the AEC to carry out some tests. I recall that they installed a radiation detector at the New York International Airport, and they caught a lady who had 200 luminous wristwatches taped to her girdle; she was duly arrested for smuggling. I am not aware of any other practical result.

Starting in the mid-1950s, I was asked to serve on numerous panels evaluating high-energy physics programs, or physics activities in general, on behalf of the National Science Foundation and other institutions. My first service, starting in 1954, was on the Physics Panel of the National Science Foundation, and I served on the High Energy Physics Commission of the International Union of Pure and Applied Physics (IUPAP) from 1958–1960. Other similar activities followed which I do not outline here.

On the military side, I was asked by Guyford Stever, who had been a fellow graduate student at Caltech and who was then Chief Scientist for the U.S. Air Force, to join a special panel of the Science Advisory Board of the Air Force starting in 1955. That panel was charged with examining in detail the possible (and impossible) defenses proposed at the time against delivery of nuclear weapons onto United States soil. The work of the panel covered both passive measures such as civil defense, and active measures such as intercept of delivery by aircraft or missiles. (The latter were just beginning to be deployed.) I found work on this panel of great interest and very educational. It demonstrated clearly that the advent of the nuclear age had drastically shifted the balance between offense and defense. Inasmuch as the delivery of even a single nuclear weapon would lead to enormous disaster, defenses now would have to meet extremely high standards in order to have significant effectiveness, and mitigation of consequences of delivery could only have limited value. This panel concluded its work with a report outlining the promise, or lack thereof, of alternate defensive measures.

Then followed my service, together with Bob Hofstadter, on an activity called "Project Metcalf." In essence, this panel, convened by the U.S. Navy, was a "shootout" between two competing air-to-air missile systems. One was the Sidewinder missile, developed by a local physics group at the Inyokern Naval Test Station at China Lake; the other, called the Sparrow, was an elaborate product of the aerospace industry. The Sidewinder was being developed in the informal World War II style by a group of physicists working on adapting an "unguided" missile to incorporate some modicum of guidance using an infrared sensor "seeker." The Sparrow was almost an order of magnitude more expensive than the Sidewinder, and aimed to develop a guided infrared homing missile "from scratch." As it turned out, both missiles, notwithstanding their disparity in cost, showed comparable promise in performance, and we reported accordingly.

In 1958, I was asked to chair a panel for the President's Science Advisory Committee (PSAC) on the detection of nuclear test explosions in outer space. This request came about as follows. As a consequence of an agreement between the Soviets and the United States, a "Conference of Experts" was convened in Geneva in 1958. This group was given the task of determining technical means of detecting and identifying a violation of a potential agreement to cease nuclear testing. If an assessment of the agreed-upon methods indicated that violations could be uncovered, political-level negotiations between instructed diplomats would be held. President Eisenhower appeared to hold the idealistic conviction that a group of scientists from countries with adversarial interests could arrive at unbiased technical assessments of problems of major international consequence.

After the 1958 Conference of Experts published its report, critics in the United States, in particular Edward Teller, maintained that the report's assessment of the power of detection was much too optimistic. Teller, together with colleagues at the Rand Corporation, set to work to invent nuclear test ban evasion scenarios that had not been considered by the Conference of Experts. Among the scenarios proposed by Teller and his colleagues was testing in outer space. In this scenario, they envisioned that the violator would send up two space vehicles, one carrying the nuclear explosive and the other carrying detection and diagnostic instruments. The mission would be programmed so the explosion would take place at a distance of perhaps hundreds of millions of kilometers from Earth, with the diagnostic satellite deploying its instruments at just the right distance from the explosion; perhaps such a test would even be hidden behind the moon! A second evasion technique they proposed was detonation of a nuclear device underground in a "Big Hole" of sufficient size such that the shockwave of the explosion, when impacting the wall of the cavity, would only produce elastic, rather than nonlinear (i.e., inelastic) distortions.

In response to these additional potential evasion scenarios, the United States approached the Soviet Union to reopen technical negotiations, and this immediately led to controversy: the Soviet negotiators maintained that the report of the Conference of Experts constituted an immutable basis for subsequent political negotiation, whereas the United States asserted that the report of the "Experts" was a technical document subject to amendment as new technical facts emerged. The Soviets eventually yielded on this point. In parallel with the drive to reconsider the report of the Conference of Experts, internal panels were constituted in the United States to analyze the newly suggested evasion schemes.

One of these was a panel on detection of nuclear explosions in outer space, which I chaired. I had not been a member of the Conference of Experts; I do not know who proposed my appointment as chair, and the members were selected by White House staff. This new panel included both Hans Bethe, who had been the senior member of the Conference of Experts, and the vocal critic of the Conference of Experts, Edward Teller. The new panel's report turned out to be unanimous, which in itself was surprising to most people, including the panel chairman. The panel set a precedent in that it did not consider any possibility of technical evasion to be determinative as such. In addition to technical evasion, it also considered the effort that the evader would have to mount in order to achieve his goal. Specifically, I became convinced that if the Soviets were really going to utilize testing of nuclear explosions in outer space as a means of evading a nuclear test ban, then the effort and cost this would incur would be so large that American national security would gain by the diversion of Soviet resources toward that effort and away from other, more nefarious endeavors.[1]

After I submitted that report, the "powers that be" decided that—because I had negotiated such a unanimous report among my colleagues—I was ready to negotiate with the Soviets. Accordingly, I was asked to chair the U.S. delegation to "Technical Working Group 1" (TWG1) which was to address evasion scenarios in outer space. I was later asked to serve as vice-chairman of TWG2, which was to address "new data" relating to the detectability of underground nuclear explosions. It was presumed that the latter group would consider not only the "Big Hole" scenario, but also additional information relevant to seismic detection that had accrued from underground testing held since the completion of the Conference of Experts report. TWG2 was chaired by James Fisk, the vice-president of the Bell Telephone Laboratories.

The call for me to chair this negotiation with the Soviet counterpart delegation came very suddenly in 1959, at a time when our family had decided to take a sabbatical leave from the work at Stanford University and, supported by a Guggenheim Fellowship, to spend one semester in Geneva at CERN, the European Center for Nuclear Research. The time sequence was such that I had to deposit my family on a boat leaving New York for Europe, and then fly over their heads to Switzerland just in time for the opening session of TWG1. Prior to our departure, I met with Secretary of State Christian Herter to receive instructions, but these were highly scanty, considering the technical nature of my mission.

Spurgeon M. Keeny, Jr., a member of the staff of the President's Science Advisory Committee and also of the National Security Council, was assigned to the Geneva delegations. He had served in 1958 with the Conference of Experts and was continuing his work with the 1959 Technical Working Groups. As it turns out, the Geneva negotiations with the Soviet scientific delegation were most educational for me in shaping my views on arms control issues. The work of the Conference of Experts and of the Technical Working Group has been described extensively in the literature[2] and complete verbatim transcripts of the negotiations are available.

The Soviet counterpart delegation was headed by Yevgeni Federov, a well-known polar explorer and geographer. Other members were equally prominent. Among them was Igor Tamm, who received the 1958 Nobel Prize in Physics. A negotiation on the political level was carried out in parallel with our technical negotiations. The U.S. political delegation was headed by James Wadsworth, and the Soviet political delegation was headed by Ambassador Semyon Tsarapkin, generally designated by our group as "Scratchy."

Our sessions were held at the Geneva Headquarters of the United Nations; they furnished simultaneous translation to the negotiators, but at the same time our delegation had its own private translator, who would monitor the official interpreters. Occasionally, our translator would slip me a note, saying something like, "Don't get mad; he didn't say that," in order to prevent a mistranslation error from escalating any controversy.

The discussions were protracted longer than had been planned. Our group met at the American Embassy after each session in order to plan the next day's dialogue. From the very beginning of the negotiation, it became clear to me that the Soviet delegation did not share our conviction that scientific and political considerations could be separated when the outcome of a dialogue between scientific negotiators could affect the interests of their country.

The interests of the two scientific delegations were obviously divergent: the Soviets were intent on minimizing any intrusive verification measures if a nuclear test ban were to be concluded, whereas the Americans wanted to attain maximum confidence in the result of verification. Typical of most differences of opinion was the Soviets' position that the detection of nuclear weapons in space was easier than had been concluded by the Americans. Despite this pattern, we reached reasonable agreement in assessing most detection methods, and the Soviets were pleasantly surprised when I introduced the possibility of using photomultipliers to detect the emission of single x-ray photons generated by a nuclear explosion in space; that method, considered earlier by our committee in Washington, resulted in the largest range of detection.

Nevertheless, some disagreements persisted. When drafting a summary assessment, I proposed to say that "The assessment is based on the current knowledge of the strength of signal and of background which we have today; future measurement may make our assessment more optimistic or pessimistic." Federov replied that dialectic materialism acts in favor of the Soviet Union, and because it is in their interest that background in the future shall become smaller, this is indeed what future scientific observations will find. Having reached an impasse on this point, we agreed to drop that whole phrase from the final document.

Another subject of disagreement concerned the use of "ionospheric radar." This is the use of radiofrequency signals reflected from the ionosphere to give an indication of any ionospheric disturbance. It was known from previous tests that nuclear explosions produced such a disturbance by their deposition of soft x-rays in the ionosphere. The Soviets objected to the inclusion of ionospheric radar as a possible detection method, and tried to introduce some highly contrived backup for their position. We concluded that the reason for this Soviet objection was probably that ionospheric radars can also detect disturbances caused by passing ballistic missiles, and the radars would therefore have military potential beyond the detection of nuclear tests. I asked for a private meeting with Federov, and maintained that considerations external to the scientific potential of detection of nuclear explosions in space were outside the province of TWG 1 and should be addressed in the political negotiations. Federov replied, "It is my responsibility to take all factors into account." So much for the separation of science and politics!

We therefore "agreed to disagree" on this point, and I cabled the State Department for their concurrence with this conclusion. I received a cable back, "What is an ionospheric radar?" I cabled in response, "Ask the President's Science Advisor." I receive a return cable saying, "The President's Science Advisor's expert on the subject is in Geneva," referring to Keeny and to me. So we cabled back saying that we agreed with this conclusion, and the State Department then cabled its agreement. This was an interesting example of the circular process of receiving instructions from the State Department on a technical subject! The report was then agreed to and signed.

The negotiations of Technical Working Group 2 were less harmonious. Two major subjects were introduced by the U.S. side. One was that "new data" showed that the seismic magnitude for a given yield of nuclear explosion would be less than that assessed by the Conference of Experts; and the second was the possibility of using a "Big Hole" nuclear weapons test in order to evade detection. On the former subject, the American position was technically weak, simply because there had been very few nuclear explosions underground, and the seismic data were of limited accuracy. Still, the disagreement remained and the Soviets, with some merit, questioned why we insisted on reopening the conclusions of the Conference of Experts.

It became the responsibility of Hans Bethe to introduce the concept of the Big Hole. Although he was an advocate of the necessity—or at least the high desirability—of concluding a nuclear test ban treaty, he agreed on physical grounds that a nuclear explosion in a cavity of sufficient size would effectively "decouple" the generation of seismic waves, and thereby decrease by a large factor its recorded seismic magnitude. All agreed that this physics conclusion was incontrovertible, however, there remained major questions of the practicality of excavating an underground cavity of sufficient size or of utilizing an existing cavity (for instance, in a salt formation) without having such a major activity detected. The Soviets appeared highly offended by our introduction of a proposal indicating how they could cheat, but after extensive discussion they had to agree on the physical possibility. The meetings were quite acrimonious, and resulted in friction between prominent Soviet and American seismologists on the two delegations who previously had enjoyed a friendly collegial relationship. Happily, these frayed relations were subsequently mended. In consequence of the controversies, the two delegations submitted separate accounts.

After the conclusion of work on TWG1 and 2, I was able to spend the balance of my sabbatical leave at CERN, working largely on the experiment on the magnetic moment of the muon, a large change from negotiating with the Soviets! After a very educational stay in Europe, our family returned to Stanford in the fall of 1959.

As is well known, protracted further deliberations between the United States and the Soviets led to the adoption of the Limited Test Ban Treaty in 1963, which banned nuclear tests in the atmosphere, underwater, and in outer space; the latter inclusion resting heavily on the work of TWG1. There then followed the agreement on the Threshold Test Ban Treaty, which banned nuclear explosions exceeding 150 kilo-tons in yield, and the agreement on banning all so-called "peaceful" nuclear explosions. A Comprehensive Test Ban Treaty was signed many years later, but it has not entered into force. Although the United States signed that treaty, its ratification was defeated in the Senate.

Concurrent with the involvement in Geneva, developments continued bearing on the fate of the proposal by Stanford University to build a two-mile electron linear accelerator. I turn to that topic later, but now discuss my further involvement in "Science Advice." As noted, I served on Sub-Panels of the Presidents Science Advisory Committee as early as 1959, and I was appointed as a member of that group in 1961 and served in that capacity until 1964.

The first chairman of PSAC, James Killian, served from November 1957 to June 1959, and wrote a book describing his experiences entitled *Sputnik, Scientists and Eisenhower*.[3] His successor was George Kistiakowsky, who served until January 1961, when, due to the change in administration from

Eisenhower to Kennedy, Jerome Wiesner assumed the office. Kistiakowsky kept a diary during his tenure as special assistant to the president that has been published in "expurgated" form under the title, *A Scientist at the White House*.[4] Wiesner did not publish such an account, but after his death, his PSAC role was fully described by his associates.[5] Thus this period, which has generally been recognized to be the heyday of science advising to the U.S. president, has been well documented, and I therefore restrict this account mainly to my own experiences and some opinions on the broad subject of science advising.

The President's Science Advisory Committee was established in the wake of public outcry over Sputnik, the Soviet satellite launched in 1957. In response to the apparent show of Soviet superiority, President Eisenhower elevated an existing science advisory body that had reported to the Office of Defense Mobilization so that it now reported directly to the president. He appointed James Killian, the president of MIT, both as chairman of the committee and as his special assistant for science and technology. My PSAC term did not overlap with Killian's, although I became acquainted with him in connection with my work in Geneva.

Kistiakowsky was an excellent chairman, and focused the agenda of PSAC tightly on questions in which the president had a direct interest or on current policy issues that had a substantial scientific component. It has often been noted that scientific input to government falls into two categories: "science in government" and "government in science." As noted, the former, that is, scientific input to policy, dominated PSAC's agenda; the latter, that is, questions of governmental support of science, were also occasionally discussed during my tenure, but were largely left to other government agencies or to outside advisory bodies, in particular to the National Academy of Sciences.

Because my period of serving on PSAC overlapped with the critical evaluation of Stanford University's proposal, submitted to the government in 1957, to fund the construction of the large linear accelerator that would become the Stanford Linear Accelerator Center (SLAC), I avoided any involvement in discussions by PSAC which touched upon that subject. At times Kistiakowsky did raise questions with me in private concerning details of the accelerator proposal.

The work of PSAC was divided into plenary sessions and subcommittee deliberations. I was primarily involved with subcommittees on ballistic missile defense, the cessation of nuclear testing, other issues related to national security, and on the manned mission to the moon. I chaired a subcommittee on the vulnerability of American missile warheads to potential enemy nuclear detonation. Specifically, we examined the impact of soft x-rays emitted by a nuclear explosion in proximity to a ballistic missile warhead. Those x-rays would cause a sudden boil-off of material from the warhead, and that boil-off would generate a mechanical shock highly destructive to the warhead. In addition, the more penetrating radiations from a nuclear explosion were expected to damage the interior electronic systems of an ICBM warhead. Along with the White

House, the General Advisory Committee of the Atomic Energy Commission was also briefed on this subcommittee's classified report.

In addition to my committee work, I participated in many of the PSAC plenary discussions. For instance, there were various deliberations on the problems of science education at the time, and a PSAC member, Jerrold Zacharias, mounted a major effort to revise school texts on science. Some environmental issues also reached the PSAC level. For instance, Congress had passed the Delaney amendment, and PSAC was asked to respond. That amendment limited the use of pesticides that may be sprayed on cranberries to such an absurdly low level that if one ate a sufficient quantity of cranberries such that the pesticides might have any remote health significance, one would be certain to expire, not from pesticide exposure, but of something else entirely!

As mentioned, under Kistiakowsky's stewardship, discussions were restricted almost entirely to fairly current topics. I. I. Rabi objected to this approach, and argued that there should be extensive deliberations on much broader science policy issues. His pressure in this regard became so irritating to Kistiakowsky that he once asked Rabi to chair the next meeting of PSAC with the sole agenda being such general considerations. Rabi accepted and the result was the most disorganized and unproductive meeting of the committee that I ever witnessed.

President Kennedy took a great interest in the role of PSAC and its chairman when he was pondering whether to send a man to the moon. PSAC examined the alternate technical approaches for achieving a manned landing on the moon, including launching a capsule either from orbit around the earth or orbit around the moon. PSAC also examined the possible scientific value of the lunar mission in relation to its expected cost. As a result, Wiesner reported to the president that the manned lunar mission could not be justified as a scientific endeavor. Instead he said that if, for political reasons, the president decided to go forward with sending a man to the moon, PSAC would help him to do a good job, provided that the president would not justify his decision publicly as serving the interests of science. Kennedy accepted that bargain, the lunar mission proceeded, and Kennedy never justified it on any basis other than demonstrating American technical prowess and therefore adding to America's prestige.

During the period of my science advising, PSAC's agenda was dominated by issues of national security. The fear generated by Sputnik was based on the perception of that event signaling Soviet dominance in respect to long-range missiles. In consequence, a great deal of PSAC's work consisted of putting into perspective initiatives by the Pentagon that promoted ill-conceived projects. Examples included such programs as a nuclear-propelled aircraft, proposed crash programs on ballistic missile defenses, the fall-out shelter program, and the like. In fact, possibly the most important result of PSAC's work during the Eisenhower administration was to render advice directly to the president on national security issues with a scientific component, rather than have that advice filtered through the parochial interests of the Department of Defense or the Atomic Energy Commission (predecessor agency to the Department of

Energy). As is well documented in Kistiakowsky's diary, this resulted in many conflicts between PSAC and its chairman and these departments.

These conflicts were particularly pronounced in respect to the continuing controversial pursuit of cessation of nuclear weapons testing. During both George Kistiakowsky's and Jerry Wiesner's tenures, pressures to resume testing were incessant, primarily from the heads of the Atomic Energy Commission and from Edward Teller and his colleagues. I was asked to participate in several committees convened to review the technical facts concerning limitations of nuclear testing.

At the end of June of 1961, I chaired a committee, composed largely of the usual culprits from earlier committees, to not only evaluate again the technical factors on the need for nuclear testing, but also to assess whether the Soviets had or had not conducted any secret nuclear tests during the moratorium on nuclear testing, which was, ostensibly, then in place. This assignment generated much publicity, and I was portrayed as a "sleuth" in the *New York Times*. On the subject of Soviet clandestine testing, our report concluded that "It was feasible for the Soviet Union to have conducted secret tests, that there was no evidence that it had done so (or had not done so), and that there was no urgent technical need for immediate resumption by the United States."[6] That judgment was not universally shared, and was criticized by the defense department.

As I noted before, the nuclear test ban controversies have persisted until this day, and although a Comprehensive Test Ban Treaty (CTBT) has been signed, it is not yet in force.

This is but one illustration of a serious problem pertaining to this and other issues of a nominally technical nature that have become objects of major political significance. At this point, the political and symbolic importance of attaining a complete ban on nuclear weapons testing substantially exceed its technical impact.

The fact is that a nonnuclear weapons state can build nuclear weapons of limited performance without nuclear testing. The Hiroshima bomb was never tested but killed over 100,000 human beings. Similarly, Israel has stockpiled a significant number of nuclear weapons without having carried out, to the best of our knowledge, a single nuclear weapons test. Thus, many advocates for a CTBT tend to overstate its technical merit in arresting the spread of nuclear weapons to states not already possessing them, or in curtailing improvements of existing weapons by states that have them. Conversely, opponents of a CTBT tend to greatly overstate the need for nuclear weapons testing in order to ensure the safety and reliability of existing nuclear weapons stockpiles, or for making "improvements" to them.

The limits of the efficacy of the CTBT in both respects are discussed in a recent National Academy of Sciences study.[7] That study foresees three possible future worlds: a world without any restriction on nuclear testing; a world where there is a CTBT obeyed by everyone; and a world where a CTBT is in existence but evaded to the extent possible without detection by the worldwide system of monitoring it provides. The net conclusion of that study's report is that U.S. national security is served better with a CTBT than without one, even if the CTBT is evaded to the extent indicated. However, the security advantages and disadvantages are not overwhelming, in particular when compared to other security threats facing the country today. The CTBT remains deeply linked politically to the nuclear nonproliferation regime.

My work on PSAC was very demanding on my schedule. The committee met in Washington on the first Monday and Tuesday of each month, and there were many additional meetings of subcommittees, or retreats of the whole committee. Because I had to teach freshman physics on Wednesday mornings, my wife would pick me up from my return flight to San Francisco on Tuesday evenings, drive to the Stanford lecture hall, and work with me to prepare the demonstrations needed for the next day's classes.. We then went home, and early on Wednesday mornings I gave the lectures and accompanying demonstrations, usually to three classes in succession.

After President Kennedy's death, the role of science advising to the president gradually deteriorated, until the science advisory bodies advising the president were entirely eliminated by President Nixon. A reason for that elimination was a conflict generated by Nixon's endorsement, against the advice of PSAC, of construction of the supersonic transport plane. Nixon and PSAC also disagreed on other issues, including ballistic missile defense. On one occasion, a PSAC member testified as an individual before a congressional committee on his negative views on the supersonic transport plane, and his testimony contradicted the president's claim that his decision was based on the best available scientific advice.

This leads me to some general observations on science advice to the highest levels of government. Indeed, science and government need each other, and many major decisions of government have a scientific or technical component. Yet the formulation of governmental policy and the advice of independent scientific bodies to a high-level governmental advisee, in particular to the president, generates a number of tensions which I list.

Conflict of Interest

A scientific advisor to government is presumed to be an "independent expert," but the problem is that he or she may be neither sufficiently expert nor sufficiently independent. Decisions taken in response to the advice given may influence the future of the advisor's field, and sometimes even the career of the advisor. Thus the advisors frequently have a direct interest in the outcome of governmental decisions.[8] Such tensions can be minimized but never fully eliminated. Good practice strives to balance the backgrounds and interests of the members of advisory bodies, but the search for balance has its limits. It clearly makes no sense to attempt to strive for "balance" between geneticists and religious creationists; the former practice a scientific discipline, the latter do not. Moreover, if highly extreme views are included among members of advisory bodies, consensus and mutually agreed-upon reports are difficult to achieve.

Who Owns the Advisor?

In parallel with independent advice sought by the president, the advisor may also be requested to furnish expert advice to other bodies, such as Congress. Although the actual nature of the advice given to the president rightfully can, and frequently should, remain confidential, once an advisor testifies as an expert individual in other forums, it can become manifest that the decision taken by the president was taken in conflict with the advice received, leading to embarrassment. I have noted how such a conflict led to President Nixon's abolishment of PSAC.

Accountability

Science advice to the president is fundamentally private, but broad science policy—and even some narrower science-policy issues—are of general public interest, and also of congressional concern. Thus Congress insisted on an accountable science advisory process, a demand that was met by the legislative creation of OST, the Office of Science and Technology, now reincarnated as OSTP, the Office of Science and Technology Policy. In addition, Congress decided in 1972 to establish a science advisory body of its own, called the Office of Technology Assessment (OTA), which was organized to include safeguards to assure nonpartisan advice. OTA was, however, abolished in 1995 by a Republican majority dissatisfied with this nonpartisan arrangement.

Access

For science advice to the highest level to be most effective, the advisor should have personal access to the target of the advice. In fact, the access of the president's science advisor to the president has been variable, depending on the president's direct interest and the personal "chemistry" between the president and his advisor.[9] Frequently, the "de facto" access has been largely limited to the vice president or to senior White House staff. Conversely, the advisor must be accessible to input from the scientific community and the relevant public or executive agencies. Thus, agencies at a lower level of government must be both free to and encouraged to communicate with the science advisor. However, the science advisor must not become a line officer through whom decisions made by the executive that have a bearing on science must first be cleared. The effectiveness of the science advisor depends on having access to the advisee. It is the potential of that access that makes the advisor's communication with lower echelons effective.

Science Advisor Versus Spokesman for Science Policy

The policymaker is free to accept or reject science advice as rendered. For that reason, tensions arise if the science advisor is used by the advisee to be an official spokesman to support the policy that is eventually decided upon. Therefore, it is best to avoid having the science advisor be a spokesman in defense of governmental policy, unless the area in question is noncontroversial, such as the support of selected scientific endeavors or general proclamations on the importance of science.

Workload

Independence of advice demands that advisors serve part-time (and are unpaid). But this imposes a limit on the number of subjects that the advisors can address, and the depth to which each topic can be analyzed. For this reason, a substantial effort of PSAC was dedicated to removing topics from its portfolio and transferring them to other bodies (such as the National Academy of Sciences) and to strengthening the scientific competence within governmental agencies.

Conflict of Advice with Preconceived Policy

Possibly the most serious tension between science advice and governmental functions arises if preconceived policy is in conflict with sound scientific advice. This has been the case in recent times in such sensitive areas as environmental policy, global warming, issues relating to reproductive health, and certain military issues where moves ostensibly in the interest of national security appear to be in violation of sound scientific criteria. Although such conflicts are unavoidable at times, their very existence points to the value—in fact, to the necessity—of independent science advice.

The science advisory system to the president still exists, but is in a state of significant disrepair, partially because the tensions listed above have frequently been resolved by allowing governmental policies to overrule scientific findings. But scientific facts cannot be coerced by policy and attempting to do so results in grave danger to the health of the country.

After my term on PSAC ended in 1964, several of my diverse advisory activities continued, but I now return to an account of the establishment of SLAC, which proceeded in parallel with Kistiakowsky's service as science advisor to the president. Some further advising activities are revisited later.

9
Establishing SLAC

It was only natural that the unquestioned success of the operation of Stanford's MARK III accelerator and the creativity of the physics research program would lead to discussions about what the "next step" should be. These speculations were initially promoted by Bob Hofstadter, who was buoyed by the success and worldwide recognition of his electron scattering program. Numerous conversations involving Hofstadter, Ed Ginzton, Leonard Schiff, myself, and some others took place at various times, and these conversations led to the presentation of a report[1] at the 1956 CERN Symposium on High Energy Accelerators examining the technical factors in extrapolating the MARK III experience to multi-GeV energies.

A series of meetings was then organized to change these speculations into reality. The first meeting, which established the ground rules for writing a proposal for a very large accelerator facility, took place in my home on the evening of April 10, 1956. All these events are documented elsewhere in much greater detail than is recounted here. In particular, after SLAC became operational in 1967, Richard Neal served as editor of "The Blue Book,"[2] which has about 90 individual authors describing all phases of the development and construction of SLAC, but which also has a chapter, written by Douglas Dupen, describing the history leading up to the submission of our proposal. Several other accounts are also given in Terman's biography and other publications.[3]

Following the initial meetings, a group was formed to produce a formal proposal, for which Bill Kirk served as the principal writer. Bill Kirk was a wonderful person. He was an English major who was recruited by Edward Ginzton as an assistant in the Microwave Laboratory, but he caught fire, developing an interest in accelerator and elementary particle physics and sharing in the excitement that the results generated. In addition to Bill Kirk's writing skills, the generation of the proposal was assisted by several volunteers outside the microwave and high-energy physics laboratories, including members of the geology department, who investigated the suitability of various sites. The actual proposal was extremely short by today's standards: a total of 64 pages in length,[4] plus several short appendices. It was separately submitted in April of 1957 to three agencies of the U.S. government, because it was unclear as to which federal agency might be the appropriate sponsor.

The Office of Naval Research (ONR) had been supporting the prior accelerator work at Stanford, but it appeared extremely dubious that ONR, as part of the Department of Defense, would expand its sponsorship to such a giant undertaking. The AEC was most experienced in supporting very large technical enterprises, but the proposed undertaking had no direct connection with "atomic energy" as such. The National Science Foundation had just been established to support basic science, but it was organizationally ill equipped to leap into construction and operation of large facilities. In view of this ambivalence in the government, the university submitted the proposal to all three agencies, and I remember conducting personal briefings describing the proposal to members of all three agencies' staffs. The White House then decided to designate the AEC as the "custodian agency" for high-energy particle physics, recognizing the experience of the AEC in that field, but at the same time signaling that the AEC was to be the caretaker of what was seen as truly a national undertaking, transcending the mission of any one agency.

In retrospect, in looking at the actual proposal, it is impressive how naïve the enterprise really was at that time. Only five pages of that proposal described the scientific mission of the two-mile machine (!). The principal thrust of our research plan was extension of the electron scattering experiments, further studies of production of existing and conjectured particles, and also speculation on the utility of secondary beams: the proposal compared estimated production of such beams by existing and projected facilities. The descriptions of the target facilities and experimental areas were extremely rudimentary, and bore little similarity to what was eventually constructed.

Considerable care, however, had been taken to construct credible cost estimates, and here we were on relatively solid ground based on past experience. We also received assistance from architect-engineering firms, which volunteered their services partially because of Stanford connections, and partially in anticipation of future involvement with the project.

After submission of the proposal, the establishment of what was initially called "Project M"—it remained ambiguous whether M stood for "Monster" or "Multi BeV Accelerator"—had to overcome many serious hurdles. The first was receiving the approval and support of the government, and successfully creating a workable division of authority between it and the university. The second was establishing the relationship of the new facility to the rest of Stanford, and the third was negotiating the relationship of the proposed "Monster" with the national community of science.

In parallel with the protracted efforts dedicated to securing project approval and overcoming administrative hurdles, technical work designed to firm up the actual specifications for building the Monster proceeded apace. Dick Neal and Ed Ginzton decided to build the MARK IV 80-foot linear accelerator underground next to the Microwave Laboratory. MARK IV served the dual purpose of a testbed for Project M components and a beam for treatment of cancer patients, thus pioneering the future extensive use of electron linear accelerators for that purpose, now a multibillion dollar industry. The building of the MARK IV facility was subsidized by funds from the General Electric Company in exchange for giving that company first but not exclusive access to the know-how developed there. These multiple missions of the MARK IV proved awkward to execute. As patients came for treatment, at times with life or death consequences, it became impossible to also manage a systematic schedule for carrying out the development of Project M components. Rather, the latter had to be done on a time-available basis. Despite this, important work that helped future design decisions was accomplished.

In addition to MARK IV, a building on campus next to the stadium was taken over for carrying out further research and development on the two-mile accelerator, and to provide offices for the project's growing staff. Next to the building, we constructed a facility dubbed "the tree house," which was in essence a short vertical section of the planned accelerator with a modulator on top of a wooden tower and a dummy accelerator section 25 feet below. We even simulated the method of assembly of the future machine by lowering a waveguide section from the top of the tower to the accelerator mock-up below. This waveguide was lowered by helicopter on a weekend just preceding a football game between Stanford University and San José State. Our employment of a helicopter generated an angry complaint from the Stanford football coach that we had been spying on the football team's practice prior to the game. We apologized.

Ed Ginzton was in charge of all this preapproval research and development work with me serving as his de facto deputy. Although these relations were never formally established, a great deal of technical progress was made.

Let me now return to the progress in overcoming the administrative and organizational hurdles enumerated above, beginning with our relations to the outside scientific community. In anticipation of the future need for presidential and congressional approval, we had an obvious interest in securing the maximum support (or at least acquiescence of) the outside scientific community. Here we were subject to diverging pressures. Some leaders of high-energy particle physics, and particularly Leon Lederman, argued that SLAC should become "a truly national laboratory," meaning the contract to build and operate the facility should be held not by Stanford University alone, but by a consortium of universities under contract with the government. In other words, Stanford should not be in any special position relative to other institutions nationally or even internationally. In response to that pressure, I contacted officials both at Caltech and the Berkeley Radiation Laboratory about the possibility of establishing a university association to manage the Monster. I was strongly rebuffed by both institutions, who indicated they had enough troubles of their own. So the "truly national laboratory" proposal did not go far.

In the other direction were the pressures originating from the Stanford physics department to have the Monster under its control, with a firm commitment for future running time being made to professors from that department. I felt that such a position was simply untenable considering the magnitude of the effort and funding required, and I concluded that such an arrangement would never be accepted nationally. The eventually accepted proposal was intermediate between a "national laboratory" and a Stanford proprietary facility. By using the term "national facility," we described in broad terms the expected relationships: SLAC would be a national facility operated under contract to the federal government by Stanford University, but that facility would be available to any member or members of the scientific community, either national or even international individuals or groups, on the basis of submitted research proposals, and on the basis of a demonstrated technical capability for execution of the proposed work. That position prevailed and was later elaborated upon through the establishment of a SLAC Program Advisory Committee reporting to the director of the laboratory, and a Scientific Policy Committee reporting to the president of the university.

Establishing relations between the proposed laboratory and the rest of the campus proved to be extremely difficult and controversial. Some of these controversies, which persisted for many years, are described in some detail in the Terman biography.[3]

At the outset, Stanford University established a University–SLAC coordinating committee comprised of the senior bureaucrats of the university and the Project M organization. In addition, the university appointed Robert Moulton, a recently arrived Stanford administrator originally assigned to deal with budgetary analyses, to serve as the university's "point man" for Project M affairs. Moulton later confessed that he had no idea what a linear accelerator was, but was initially too shy to request an explanation.

Agreement was reached that the new laboratory should be a separate administrative unit of Stanford University, rather than an adjunct to any one department or school. However, that decision left unanswered many questions that were directly related to maximizing the expectation for both successful design and construction of the facility and for its operation for creative research. I had given careful consideration to the modality of carrying out the future research program of the laboratory, and came to the conclusion that the techniques for doing successful experiments required lead times for constructing experimental facilities and equipment that would be just as long as those needed for constructing the basic accelerator itself.

This conclusion differed sharply with existing practice at other accelerators. At past proton accelerators, the machine was completed, and afterwards experimentalists "scrambled" to assemble experiments from existing components, or to build such components rapidly from scratch. This certainly had been the case at Berkeley, and was also the case at CERN when their first machines started operation.

But I concluded that such an approach was clearly impossible at a high-intensity electron accelerator operating at a short duty cycle, that is, producing a pulsed beam of duration of the order of a microsecond at a rate of several hundred times per second. This implied that conducting traditional experiments using detection based on coincidence in time of correlated events would be extremely difficult because accidental coincidences would pile up during the short beam pulse. Moreover, when a high-intensity electron beam is stopped, the result is an electromagnetic cascade with a high multiplicity of particles; this, too, would cause serious background problems. Therefore, the design of successful experiments would require building extensive specialized equipment. In turn, the creation of such an experimental facility demanded talent and leadership in experimental high-energy physics which, in contrast to the talent available in accelerator and microwave technology, we had not as yet assembled at Stanford.

All this implied that Project M, later to become SLAC, should establish a staff of senior physicists who would be its intellectual leaders, both for constructing experimental facilities and for conducting some of Stanford's share of projected research on the machine. The prospect that outside users could fulfill this function was minimal. At that time, high-energy physics was growing all over the country, and although there was very little opposition within the physics community to the creation of SLAC, the general attitude was acquiescence rather than enthusiastic support. Thus, outside user participation, although expected, was anticipated to grow slowly, and could not satisfy the immediate needs.

One possible organizational structure considered was to have no distinction of any kind among members of the scientific staff; a model for that approach was the practice then existing at the Bell Telephone Laboratories, where all professional technical employees were "members of the technical staff." But such a "classless" structure was completely incompatible with that existing in the university or elsewhere in the high-energy physics community. Within the university, the professoriate constituted the acknowledged intellectual leadership and I decided that in view of all the needs as visualized, it was necessary to establish a SLAC faculty. This proposal was accepted by the provost and president but was strongly resisted by the physics department. There ensued a long-lasting controversy.

The physics department reluctantly accepted the existence of a Project M faculty, but proposed to enjoin the members of that faculty from supervising graduate students of the physics department working toward a Ph.D. degree unless exceptions to that policy were approved well in advance. Moreover, SLAC faculty members were to be "prohibited" from teaching courses, and the SLAC faculty appointments would be "coterminous." This latter designation was to mean that tenure would constitute a first lien on any SLAC resources but would not extend beyond the existence of SLAC. I note that this is actually no different from the regular tenure in university departments: if a department is abolished, the tenured professors do not have a right to a continued position in other departments. This policy was put to the test in the past when Stanford University abolished its department of architecture. In addition to these restrictions, the physics department also took a strong stand against any joint appointments. However, members of the SLAC faculty were to be regular members of the academic council of the university, serve on university committees, and enjoy other such dubious privileges.

The controversy about joint appointments resurfaced several years later when, with the help of John McCarthy of the computer science department, SLAC searched for a head of its rapidly growing computational activities. The leading choice appeared to be William Miller (who later became the provost of Stanford University). The computer science division of Stanford University (later to become a department) and SLAC proposed a joint appointment, which appeared to be a logical move. However, the physics department, which was not directly involved, voiced strong objections, and even solicited letters of support from other departments. The administration reacted quite angrily, and the joint appointment proceeded.

I note here that all these matters have now been largely forgotten or settled constructively and peacefully. Relations between SLAC and the physics department are amiable; there are in fact some joint appointments and a great deal of cooperation exists in many other respects. SLAC faculty is solicited regularly to fill vacant teaching assignments in the physics department, but SLAC faculty members do not have course teaching obligations.

In summary, the basic structure of the new laboratory within the university was designed to be "academically joint, administratively separate." The latter provision meant that the new laboratory would contain its own infrastructure, including administrative departments and would retain responsibility for maintenance of its own grounds and facilities. This made sense, inasmuch as these activities would be sufficiently large to be efficient as autonomous operations. It would also mean that administrative costs would be direct charges to the envisioned contract with the government; therefore Stanford indirect costs, whose magnitude had been a controversial issue in the rest of the university, would be only a few percent, with SLAC covering such small items as the burden SLAC might generate on the top administrative offices of the university.

In parallel with establishing this new structure and proceeding with some preliminary recruiting, the principal remaining issue was of course to solicit government approval of the proposed project. Again, this turned out to be quite controversial, but I note that the time interval between proposal and construction approval (about five years) was short compared to current experience.

An essential component of securing government approval was selection of an appropriate site. Three different alignments for the accelerator on Stanford property were considered: the surviving location was the present route paralleling Sand Hill Road; the other sites were rejected because of geological problems. A site near the Bay and Moffett Field was rejected because it would liquefy under earthquake conditions.

The present site does not cross any active fault lines, although the injector is only at a distance of one-half mile from the San Andreas rift zone. Approximate boundaries of the proposed site, comprising over 400 acres, were established. One controversial issue was whether Stanford University should charge rent to the government as part of a projected contract between the government and the university. This issue came up in the congressional hearings, and Moulton assured congressional staffers and congressmen—without having first secured firm authority for that assurance—that no rent would be charged. When even the possibility of rent had been mentioned, it had caused considerable ire on the part of several members of Congress. Following Moulton's somewhat premature commitment, the issue was settled by Terman, who sharply questioned the board of trustees as to whether Stanford was an academic institution whose mission was research and teaching, or a real estate enterprise; this argument prevailed in persuading the balance of the administration and the trustees to settle for a token $1 per year lease provision.

In its efforts to secure U.S. government approval, SLAC faced opposition both within the executive branch and Congress. The situation within the executive branch is well documented in George Kistiakowsky's diary.[5] John McCone, the chairman of AEC, was very unenthusiastic about supporting SLAC, and he made quite a few moves to block its approval. A special committee was appointed jointly by the President's Science Advisory Committee and the General Advisory Committee to the Atomic Energy Commission (dubbed GAC-SAC, chaired by Mannie Piore) and that committee gave a very favorable recommendation. Subsequently, President Eisenhower gave a speech before the American Association for the Advancement of Science and others[6] in which he explicitly endorsed funding for construction of an electron linear accelerator at Stanford.

Nevertheless, sniping within the executive branch continued. I. I. Rabi intervened, proposing that an intense neutron source be constructed instead in the East. Kistiakowsky commented on that proposal with the remark, "What a bastard." McCone also raised the issue of potential conflict of interest with Varian Associates, noting that Ed Ginzton had a dual interest both in SLAC and that company. Eugene Wigner raised the interesting objection that SLAC would draw so much physics manpower that there would be insufficient talent available for military research. Kistiakowsky, being faithful to the commitments made by the president, repeatedly opposed many of these initiatives.

Requests for authorization of the proposal went before Congress for fiscal year 1960, and hearings began before the Joint Committee on Atomic Energy (JCAE)[7] in 1959. At those hearings, many statements in support of the Stanford project from senior scientists were heard; an independent cost analysis of future operating costs was submitted; and a great deal of technical material was provided. However, the JCAE and its chairman (Clinton Anderson) were miffed about the fact that President Eisenhower, a Republican, had announced his support for SLAC before having consulted the Democratic JCAE. Accordingly, the committee severely grilled Ed Ginzton, who was testifying on behalf of SLAC (I was in Geneva). Chairman Anderson raised the conflict of interest issue, the earthquake risk to the accelerator, and several other matters. Questions of cost uncertainties were also raised. There was a tense moment when a congressman from the State of Washington proposed that the SLAC accelerator be located in an abandoned railroad tunnel in his state. An AEC witness rejected that proposal on the incorrect grounds that the accelerator alignment had to be exactly level. This testimony gave me a worry later when we decided to slope the accelerator downhill by 0.5% (50 feet in two miles) in order to reduce earth-moving costs.

The JCAE deferred authorization at that time and instead recommended allocating an amount of three million dollars for continuing research and development for the project. An attempt to restore full authorization in the House of Representatives failed. I note that the matter of conflict of interest was definitively settled by a later formal agreement among Varian, Stanford, and the AEC that Varian would not be a candidate for bidding for klystron tubes and other components for SLAC for a specified number of years.

On September 15, 1961, in the next fiscal year, the JCAE relented by jointly approving authorization of construction of SLAC (a "Republican" project) together with a project to adapt a Hanford plutonium production reactor to produce electricity from its spare heat (a "Democratic" initiative).[8] Unfortunately, the latter project was in violation of the second law of thermodynamics, in that the effluent temperature of the Hanford reactor was so low that it precluded efficient generation of electricity. The Hanford project was subsequently canceled. Nevertheless, the Stanford–Hanford compromise gave birth to SLAC, and I always maintained that the real reason for that compromise was that it rhymed.

During the JCAE hearings, the estimated construction cost for SLAC was fixed at $114 million, of which about 25% was identified as escalation and contingency; the former to provide for anticipated inflation during the construction period, and the latter providing for unforeseen requirements. In addition, Congress appropriated $18 million for providing research equipment, in accordance with the foreseen need for such equipment as discussed above. Support was also continued separately for continued research, development, and operation of completed facilities.

These very reasonable provisions were generated both as a result of the hearings, but also in response to the quiet work of AEC staff with the staff of the JCAE. In particular, Herbert F. Kinney, the special assistant to Paul McDaniel, the director of research within AEC, played a constructive role. Kinney clearly identified the interests of all parties, the JCAE, the AEC, and Stanford University, and he helped to get the project moving in a spirit of mutual partnership. SLAC owes him a great deal of thanks.

A first contract for the R&D and architect-engineer work was negotiated in late 1960. Stanford University and the AEC appointed negotiating teams to agree on a contract. I chaired the Stanford contingent. Most of these negotiations proceeded in a fairly routine manner, establishing the responsibility of the university to manage the program of SLAC, and defining the AEC's responsibility to oversee Stanford's activities. But a major controversy was generated by disagreement over the division of responsibility for construction of the "conventional facilities" between the AEC and Stanford University.

The AEC negotiators insisted that the architect-engineer contract for construction of conventional facilities, that is, buildings, site preparation, and construction of the housing for the accelerator and experimental facilities should be directly contracted by the AEC, because presumably Stanford lacked experience in this area. At the same time, the responsibility for the design and construction of the technical components, the accelerator itself, and the research facilities, would remain with the university. The AEC cited some experiences at the Argonne National Laboratory where some significant cost overruns had occurred during construction. I considered this division of responsibility to be highly impractical, because there were innumerable details where the construction of the conventional facilities and technical facilities would interact. Therefore, having to involve the AEC staff each time some hole where a wave guide would run through a concrete wall had to be moved, or when some other modifications of conventional construction were necessitated by the technical arrangements, would be time consuming at best and impossible at worst.

The negotiating teams could not agree on this point, and the AEC team said that they were instructed from Washington not to accept unified construction responsibility for the university. We agreed to arrange a "peace conference" on Alameda Island. The Chairman of the AEC, John McCone, and General Manager of the AEC, General Luedecke, flew in accompanied by AEC staff members. On the Stanford side, David Packard, who was chairman of the board of trustees at the time, and President Sterling participated as well as myself and Richard Neal. At the start of the meeting, General Luedecke and I each stated our positions. Then David Packard, whose company at that time probably controlled more money than that necessary to build SLAC, was asked for his opinion. He said in a low voice, "Well, we should accept the position of the people who are going to do the work." And John McCone thought this was very reasonable, so the matter was settled; Stanford University retained unified responsibility for all construction activities. Once this "cliffhanger" was resolved, the contract for R&D and architect-engineer work with the university was signed, and work could proceed.

A second, protracted contract negotiation started after Congress finally authorized construction in September 1961. Negotiations again proceeded routinely at first, defining the university's overall program responsibilities while giving the AEC the right to approve expenditures exceeding a certain amount, and also to approve the appointment of a university-selected director of SLAC. Although it was the university's intent to nominate Ed Ginzton for that post, the nomination did not materialize for two reasons. First, Ed Ginzton had been heavily attacked during the congressional hearings on the basis of the alleged conflict of interest issue, although in my view this was a pretext for delaying authorization. In addition, the last of the Varian brothers had died, and Ginzton was now the only survivor of the original group that had established Varian Associates. Because of this, Ed felt he could no longer straddle the academic and industrial worlds. He accordingly took over leadership of Varian Associates and resigned from the university.

As a result of this situation, I became director by default and served in that capacity for over 20 years. Interestingly enough, we have been unable to find any record—either in the archives of the board of trustees or of the president's office—that I was ever officially appointed. The President of Stanford, J. E. Wallace Sterling, was somewhat concerned by Ed's resignation as director, and prevailed upon Ed to serve as a consultant to him throughout the construction period of SLAC in case there were substantial difficulties. That consulting never happened.

During this second round of contract negotiations, it was also agreed that Stanford University policies, to the extent they were applicable to SLAC, should pre-empt AEC or other governmental policies; this provision was of particular importance in respect to personnel policies, including wages and salaries. At the same time, the SLAC director held a position within the Stanford bureaucracy equivalent to a university vice-president; in that function he served on the President's Council, which held regular meetings. Thus although AEC was to abide by relevant university policies, the SLAC director would have a voice in formulating those policies.

The issue of the relation of SLAC to the wider scientific community was also settled, through the contractual establishment of a Scientific Policy Committee (SPC), reporting to the president of the university. The president was to transmit that portion of the SPC's report addressing SLAC–outside user relations to the AEC.

After agreeing during the early negotiations on basic principles on the AEC–university relations, a number of very serious conflicts arose, some of which endangered the very existence of the newly authorized laboratory. The AEC asked for a provision that the university would dismiss any employee whose continued employment, in the view of the AEC, was "not in the National Interest." I resisted this very strongly, and the AEC negotiators yielded on that point.

But then the AEC insisted on a provision that would require Stanford to undertake any work at SLAC, again "in the National Interest," which the AEC deemed necessary: this provision was intended to compel the university to have SLAC undertake classified military work when demanded to do so by the AEC. I considered this to be unacceptable and in a personal conversation told Glenn Seaborg, who had succeeded John McCone as chairman of the AEC, that this would be an impossible condition for the university to accept. Seaborg told me that he agreed with me, but that he would be outvoted four to one within the AEC if the matter escalated to a commission vote. I insisted that the AEC should indeed consider the issue, and Seaborg prevailed on a vote of five to nothing. It is noteworthy that although the other university negotiators sympathized with my position, they did not take as firm a stand as I did.

The contract was finally signed in April 1962. A separate 50-year lease was signed as well, giving the United States full use of the Stanford acreage at one dollar a year, provided the university could participate in SLAC's intellectual endeavors to a significant (defined in the lease) extent.

Ground was officially broken to start SLAC construction in July 1962. Many U.S. dignitaries attended, including the presidential science advisor. When looking at the site, Felix Bloch from the physics department said to me: "Pief, if you must build a monster, build a good monster," an admonition I accepted.

The foregoing has dwelt on the hurdles that had to be overcome before final authorization was granted and before agreements were reached as to how SLAC should relate to the national community of high-energy physics and the Stanford community. Of course, actual appropriations to match the authorization would have to be made each subsequent year by Congress. But it is hard now to appreciate the elation that followed the congressional action which fully authorized SLAC's construction and scientific future.

10
Building a Laboratory

Establishing SLAC transcended construction of the accelerator complex, research facilities, and buildings: it meant creating a total environment that would enable great people to do successful work. This is obviously not the place to give again an account of the construction of the SLAC Laboratory. As noted previously, that has been done fully in the very comprehensive "Blue Book" that Richard Neal edited.[1] I wrote some of the material in "The Blue Book's" chapter on beam dynamics, as well as editing the rest of that chapter. I describe here only some of the general guiding principles for managing this new enterprise, principles that we employed to maximize the chance of success of the new laboratory. I also describe some of the activities in which I participated as a scientist, rather than as the director of the laboratory.

As noted above, we concluded that SLAC needed a faculty to establish intellectual leadership for the technical, scientific, and educational activities of the laboratory, and we proceeded to recruit that faculty without delay. However, it was clear from the very beginning that the success of the lab depended on all its members, whether they were blue-collar workers, technicians, administrators at all levels, or scientists. Each person should identify his or her personal success with the success of the laboratory as a whole, and this in turn would require that all members of the staff be kept as informed as possible of what was going on, not only in respect to their immediate responsibility, but to the laboratory as a whole.

In accordance with this principle, the director's office maintained accessibility through an open-door policy. Any member of the staff could drop in at any time. That opportunity was not abused, and it proved to be very useful and helped to maintain everyone's interest. Inversely, I did a lot of running around covering all areas of the laboratory, including carrying out so-called "safety walk-throughs." For these I was usually joined by SLAC's one-man safety department, the safety officer by the name of Fred Peregoy. We would chat with people in the shops or with construction crews. If any safety issues were identified, Peregoy would note them; specifically if unsafe equipment was identified, Peregoy would tag it and procure a replacement.

On one of my walk-throughs, I opened a door to the klystron gallery and was promptly hit on the head by a light I-beam that had been leaning against the door. I wrote a memo to Fred Hall, then head of the plant engineering department, saying that I considered such an event "downright inhospitable." That memo induced an unprecedented clean-up of the site, much more thorough than would have resulted had I directly requested such a house-cleaning. In addition to these walk-throughs, we had many "all-hands" meetings to discuss progress, and we tried hard to avoid administrative jargon in communications to the technical staff.

We established an extremely simple line organization. In addition to the director and deputy director, there were associate directors for administrative services, business services, the research division, and the technical division. We specifically used the term "services" to designate the first two divisions to signal that the end product of the Laboratory was science and technical achievement, rather than "law and order" in the conduct of administration or business. We also made sure that these four divisions were organized along the lines of "what" is to be accomplished, rather than on the basis of "how" to accomplish it. Among other things, this strict line organization was to signal that safety and accountability for resources was a line responsibility that should not be diluted by establishing separate "boxes" for "how to" functions. Figure 10.1, taken

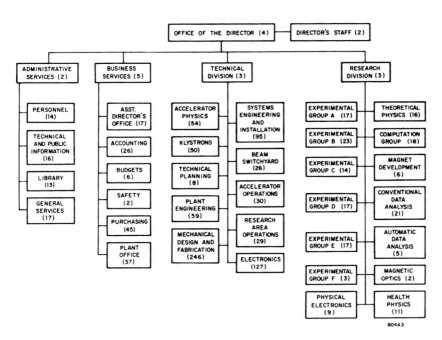

Figure 10.1. Organization of SLAC during construction and initial operation. (From SLAC Archives, Panofsky papers.)

from "The Blue Book," illustrates the organization, which was not changed throughout the construction period of SLAC and for many years thereafter.

This simple organization caused some criticism at various times from the AEC and its successor agencies. Whenever there were problems in some part of the AEC structure with issues such as safety, operational efficiency, control of capital equipment, and similar problems, the AEC tended to request that we should establish a special officer, reporting directly to the director, to be in charge of making sure that each of these functions was being responsibly executed. I resisted such pressures because, as a practical matter, they would have diluted the responsibility of those having line responsibility in doing their jobs safely and efficiently. This matter came to a head once when the manager of the San Francisco Operations Office called me into his office and demanded, "Panofsky, you have to increase your administrative functions even if it decreases the research output of the Laboratory." As I recall, my answer to this request was, "You're kidding!" I reminded the manager that, ultimately, the taxpayer was paying for research output. During those days, this kind of frank discussion was entirely feasible, because, rather than a formal government-to-contractor relationship, there was a real spirit of partnership between the AEC and the laboratory, despite all these altercations. This informality extended throughout all technical and scientific interactions between the AEC and laboratory staff: in the fiscal area; the oversight functions of the AEC were generally more formal.

One critical issue I faced was to stay involved in technical matters despite the press of administrative necessities. Thus I attended technical meetings at many levels of operation, but made sure when I participated in technical discussions that none of my remarks might be interpreted as my making decisions for others, or my overruling the authority of the person who was running the meeting. I attended technical meetings as a staff member among his peers and friends, and this was generally understood.

A further matter deserving a great deal of attention was the relationship of SLAC with industry. In respect to conventional construction, we negotiated a prime contract with a joint venture of architects, engineers, and constructors, called ABA, which stood for "Aetron–Blume–Atkinson." The details of this arrangement and the excellent qualifications of the contractors are fully described in "The Blue Book." Otherwise, we strove to build up sufficient in-house facilities so that, in case outside contractors did not perform well, we could pull the work back into the laboratory for at least small quantity production. In other words, we adopted the principle that, wherever possible, we would only ask industry to do things that we could also do in-house in small quantity. That principle served us well in many instances.

The above account is just an outline of guiding principles, and of course the success of the laboratory depended critically on the quality of the people we were able to attract. When making appointments for associate directors, department heads, and group leaders in technical and scientific areas, we tried very hard to select such key individuals to be technical people first and administrators second. This general priority of choice served the laboratory extremely well.

In 1960, Richard Neal was appointed associate director in charge of the technical division. In that capacity; he was what today would be called project manager for the construction of the laboratory facilities. He did a fantastic job in systematically managing all construction activities. He created a technical planning department directly reporting to him that established "critical path management" (CPM) methods for keeping track of progress and expenditures. Neal would meet with each of the group leaders at 8 AM each workday, so that he would cover the work of all the groups in about a two-week period; during each of these meetings, the CPM network would be updated as it pertained to the particular activity. Neal would then conduct a weekly technical division meeting, which I attended, where the progress of each group was discussed in turn. Due to this highly systematic approach, which Dick Neal attributed to his U.S. Navy background, everyone was kept fully informed about how each activity was keeping up with the overall schedule and how much delay was tolerable in each activity without endangering adherence to the general timetable.

Joseph Ballam was appointed associate director to lead the research division in 1963, and held that job for 19 years. He was a wonderful person. Although his own research concentrated primarily on hydrogen bubble chambers, he was able to convey a total sense of evenhandedness in all program decisions, so that even those disappointed by such decisions were satisfied in their fairness. He came to the SLAC faculty in 1961 from Michigan State, and before that, he had been at Princeton University. I had met him earlier in his career, when he was working at Berkeley on cosmic ray physics.

The business services division was led by Fred Pindar, who transferred to SLAC from the Hansen Laboratories on campus. He established an excellent division, which conducted its work without drawing any significant criticism from AEC, a remarkable achievement. The administrative services division was directed by Robert Moulton, who had played a vital role during the Congressional authorization process described in Chapter 9.

Our first deputy director was Matthew Sands, whom we recruited from his work at the Caltech electron synchrotron. He had achieved an excellent reputation during the war at the MIT Radiation Laboratory, and had produced one of the major reports on the electronic advances at that facility when its existence became public. He also had worked with Richard Feynman at Caltech on producing the famed "Feynman Lectures." Sands was an individual dedicated to maintaining the highest possible standards, a difficult job when dealing with an extremely diverse community such as the one existing at SLAC. He left SLAC in the late 1960s, and subsequently assumed a position as a professor at UC Santa Cruz, but has continued to collaborate with SLAC on its research program.

The construction of SLAC was completed in 1967 on budget, on schedule, and exceeding its originally proposed performance. This record, which then and also today is very rare in the government-supported defense industries and also in reactor construction, has left us with a very uncomfortable halo which thus far has not been shed.

One reason for the successful budget control was that we enjoyed a great deal of latitude or "freedom to mix" among the different budget line items. As a result, the actual construction of the accelerator and its housing was accomplished essentially within the projected budget, and most of the contingency dedicated to the accelerator was used for the completion of those other items, in particular those in the experimental target areas and the beam distribution system, where our experience had been marginal at the time the proposal was submitted.

Let me give one example of cost reduction: one of the largest budgetary lines was the construction of the modulators to supply pulsed power to the klystrons. We initially requested bids from industry to supply instruments meeting our specifications for performance. In reply, we received unsatisfactory and expensive proposals, and we rejected them all. Instead, we decided to take responsibility for performance in accordance with our needs, and we produced and documented well-working prototypes in-house. We then requested bids on the basis of "build to print" performance, rather than performance specifications. The awarded contract resulted in a saving in cost by a factor of two relative to our estimated budget costs.

Another cost-saving move was to cancel the construction of a separate prototype small accelerator to test accelerator components. Rather, we placed construction of the first two sectors of the actual two-mile accelerator onto a crash construction schedule, so that those first two sections could serve as testbeds for accelerator components. Our experience with the earlier accelerators at Stanford made the risks of this move acceptable.

A great deal of the contingency went into the construction of the so-called beam switchyard under the general direction of Richard Taylor. That switchyard incorporated provisions for three independent beam lines, rather than the two envisaged in the proposal. Moreover, it was constructed very conservatively in regard to radiation containment and risk reduction, in case any leak of cooling water or other malfunction occurred. The beam switchyard incorporated a stainless steel catchbasin in its foundations to prevent seepage of any possible effluents from reaching beyond the building. Also, provisions were made for remote assembly and disassembly in the switch-yard of elements that would become highly radioactive. None of these features had been anticipated in the original proposal. Moreover, we provided a great deal more facilities and office space for experimental users than had been proposed at the time of authorization.

The above examples illustrate that the "freedom to mix" among budget lines actually served containment of costs rather than encouraging unforeseen expenditures. This lesson is often forgotten when the government imposes line-item budgetary controls on its contractors.

With the directorial team in place and after intensive further recruitment, construction proceeded at a brisk pace. At this point in the organization's history, there were many technical decisions made, and I now highlight some instances for which I recall some personal involvement. One of the first decisions had to address construction of the accelerating structure. As mentioned previously, the method of assembly invented by Hansen to expand the disks within that structure, and in so doing to produce a shrink-fit with the outer cylinder, proved unsatisfactory. We pursued two methods of fabrication: the first was to "electroform" the structure by plating the outer cylinder onto an assembly of premachined copper disks separated by aluminum spacers that were later dissolved in sodium hydroxide. The second approach, which we adopted, was to braze premachined disks and rings together to constitute the accelerator assembly. This latter fabrication method had the advantage over the electroforming method in that the fabrication time for each section was sufficiently short, so that errors could be more quickly discovered and corrected.

Arnold Eldridge was in charge of fabrication of accelerator components and did an amazing job in setting up production of the accelerator sections. We had decided to use the constant gradient structure as designed in detail by Greg Loew, rather than what was known as the constant impedance configuration that had been adopted in earlier machines. In the constant gradient structure, each successive small cavity incorporated an iris of decreasing size, such that the group velocity of the wave propagating into the structure was diminished as the wave propagated. In consequence, as the energy of the wave became depleted, the filling time of each successive cavity was increased, so that the accelerating fields could build up to the same gradient. This meant that the structure had to be assembled such that each ten-foot section incorporated disks of different dimensions, in order to meet the constant gradient objective.

Eldridge assembled a team of half-time workers, mainly women who had time available from domestic responsibilities. These part-time workers assembled rings and disks in precise assembly fixtures made of granite, and then brazing proceeded after a thin layer of eutectic silver solder was placed between the rings and disks. Roughly 100,000 brazed joints were made by this method, and none of them have leaked in over 40 years. Each assembled section was carefully tested for its radiofrequency characteristics using methods largely developed by Gregory Loew. The accelerator sections were assembled on "strongbacks" consisting of large-diameter rigid aluminum tubes that supported and stabilized the more pliable copper structures. The design and assembly of the total accelerator structure is described in detail in "The Blue Book."

As construction progressed, I became interested in several related detailed technical subjects. One was the method of precision alignment of the strongbacks on which the accelerator sections had been assembled, and another was the problem of monitoring the radiation that would be emitted in case the accelerated beam were lost somewhere along the accelerator structure.

Conventional survey methods were used to map the SLAC site and a reference baseline was established between a tower at the injector end of the accelerator and one erected on the hill behind the experimental area at the accelerator's opposite end. However, the accelerator itself demanded more precise alignment than could be achieved with conventional survey methods of accuracy limited by atmospheric irregularities. I concluded that the alignment of the accelerator should establish that the machine was straight, but that the precise direction in which it pointed was irrelevant. We decided therefore to use a point-to-point base alignment system.

The solution chosen was to install a laser on the east end of the accelerator and then, every 40 feet inside the strongback sections, install Fresnel diffraction screens whose diffraction pattern would generate a precise crosshair on the west injector end. The strongback cylinders would be connected to each other by sliding pins inside bellows, and evacuated to a rough vacuum near 10^{-4} mm. The Fresnel screens were computed and built using the skill of several physicists and engineers;[2] the precise pattern chosen is described in "The Blue Book," as is the method of determining the exact positioning of the screens relative to the accelerator axis. The system works well; readjustment of the accelerator is required very infrequently (but was critical to restoring alignment after the 1989 Loma Prieta earthquake).

Another item of interest to me was the determination of where all or part of the accelerator beam might be lost along its two-mile stretch. I suggested using a single air-core coaxial cable stretched along the entire two-mile length of the machine to serve as an ion chamber. If and when the beam were lost, the timing of ionization within the cable would indicate the location of the loss within 100 to 200 feet, and the existence of such ionization, if of sufficient strength, could serve as an alarm to trigger beam shutdown in time to prevent damage to the structure. Because the beam and the signal along the coaxial line travel at the same speed, the reflected signal must be used to define the timing. The system was installed, and works fine. In addition, beam position monitors at the end of each 333-foot section permit precise location of the beam to within 1 millimeter.

In connection with the foregoing, let me make some comments on radiation protection at SLAC. When formulating general policies for the laboratory, it was clear that permitted radiation tolerances had been steadily decreasing; therefore, after extrapolating that decrease over the projected life of the accelerator, I established radiation tolerances for SLAC that were considerably lower than those established by the AEC for workers potentially exposed to radiation at its laboratories. These lower tolerances caused some nervousness at other AEC establishments, and I was asked to write a letter to the AEC certifying that our tolerances were "administrative" and not dictated by physical or medical necessity (!). Such a letter was written.

In preparing the initial proposal, I did some of the calculation to specify the required radiation shielding for the target areas and to establish the distance between the belowground limited-access accelerator housing and the aboveground, continually accessible klystron gallery. After SLAC was established, a strong radiation physics group, under the leadership of

Richard McCall, performed a great deal of analysis on radiation shielding, residual radioactivity, potential neutron skyshine, and associated issues.

We concluded that it would be a good idea to load the belowground concrete accelerator housing with boron around positions of potential beam loss; it is well known that a boron isotope has a large neutron capture cross-section, which leads to stable isotopes, thus preventing residual radioactivity. A search for suitable boron minerals suggested that colemanite, a crystalline boron compound, might be suitable to be added to the concrete aggregate. My wife and I had a wonderful time going to Death Valley and scrambling underground in a colemanite mine that was then operating there. We collected lots of samples. After that trip, the lab procured a whole trainload of colemanite, which was added to the aggregate poured around the positron source, and around other sources of beam interception such as slits and collimators. (Incidentally, the colemanite mine in Death Valley is no longer in operation.)

Let me turn to the problem of electric power supply to SLAC. The original intention was to run a dual circuit, that is, a six-conductor power line operating at 220 kilovolts as a branch from an existing line along the skyline on top of the hills east of the laboratory. A secondary line at 60 kilovolts was to be run from a substation on the main Stanford University campus to serve as backup in case of breakdown or maintenance on the main line. At that time, the neighboring town of Woodside had been sensitized to the visual impact of power lines strung on conventional steel lattice towers; in particular, the skyline circuit to which we planned to connect had a large visual impact on the community.

When we made our plans known, a great deal of protest ensued. Accordingly, we reduced our goals by changing our line to a single circuit, and with the help of a consultant, H. Halperin, replaced the conventional tall steel lattice towers with hollow steel poles of approximately one-half the height of the towers. This reduced height was made possible by having the porcelain insulators cantilevered sideways rather than hanging down, as was the case in the conventional structures. Moreover, we arranged to have the line strung under high tension, thus reducing the sag between the poles. This was achieved by first placing pulleys on the insulators which, by equalizing the tension between the two line segments, would permit increased tension. In addition, we planned to string the line by helicopter, which would allow us to avoid establishing access roads for construction.

This design, resulting in a greatly reduced profile and impact relative to the conventional high-voltage power lines, did not satisfy the citizens of Woodside, who demanded that the line be put underground. We studied that alternative and found it exceedingly difficult due to the elevation difference of several thousand feet along the branch line. At that time, underground power lines carrying voltages as high as 220 KV consisted of paper-insulated cables in a pipe carrying high-pressure oil. Such underground lines would not work over large differences in elevation, so our planned route would require several substations along the way. This alternative would have been very much more expensive and still unsightly, and we therefore did not accept this option. As a result, the conflict with Woodside became more intense, and the city hired a lawyer, Pete McCloskey (who later ran successfully for Congress).

The controversy reached the White House and President Johnson asked the Commission on Natural Beauty, chaired by Lady Bird Johnson, to investigate the matter. In turn, that Commission deputized Laurance Rockefeller to make an inspection of the local situation. Laurance Rockefeller appeared on the SLAC site in the longest limousine I had ever seen, and encountered Doug Dupen, SLAC's director of technical and public information. He announced himself by saying, "My name is Laurance Rockefeller and I have come to make an anonymous inspection of the power line situation." Dupen was highly astonished, but immediately gave him a tour of the proposed power line alignment and the city of Woodside. Rockefeller returned to Washington and made the Solomonic recommendation that SLAC would put its power line underground once Woodside had done the same with their own power lines. Rockefeller had correctly observed that the city of Woodside was graced with an extensive network of aboveground lower-voltage power lines which were at least as tall as the pole line proposed in the new SLAC design. I was also asked to debate with the mayor of Woodside on public television in a forum presided over by Caspar Weinberger, and I defended our low-profile design.

The city of Woodside went to court, pointing out that a section of the Atomic Energy Act enjoined the AEC from engaging in the transmission of electricity. That provision was established in connection with the AEC generating electricity at its reactors, not in connection with the AEC being a consumer of electricity. Woodside won. As a result, I undertook a lobbying campaign in Washington, resulting in an amendment of the Atomic Energy Act, restricting the applicability of the provision on the transmission of electricity to the AEC generating electricity. As a result, AEC was able to obtain the power line right of way, and the line for SLAC was built. I recall that a delegation of Woodsiders came to inspect the new line and proposed that five of the poles should each be named for an AEC Commissioner. An unnamed AEC staff member who was present said, referring to one of the commissioners, "We would have to find a pole that bends."

A final remark on the power line epic: at the very end, when power was to be connected to the line, the Pacific Gas and Electric Company engineers refused to connect to it, claiming it was substandard. I asked why, and they said that at some of the poles, the connecting power lines were joined at an angle. I asked how large an angle would be considered permissible, and received the answer, "Zero." So much for quantitative engineering. PG&E ultimately agreed, power was connected, and the line remains in use today.

The supply of power, and in particular, the cost of power is a continuing issue. After the power line connections were established, a contract was negotiated under which SLAC was supplied by a combination of "socialist" and "capitalist" power. A certain block of power was allocated to SLAC from public power through what is now the Western Power Administration, with the local power company (PG&E) receiving a "wheeling" charge (as if electricity ran on wheels). The balance of SLAC's demand would be supplied by PG&E. I secured an interview with the chairman of the board of PG&E, pointing out that SLAC, with its large load, near 100 MW, did not receive enough support to run continuously. Therefore, we would be willing to interrupt our demand at any point in time where the total load on the power system might threaten to become excessive. We would accept such an arrangement for interruptible power in exchange for a lower rate. I received the answer: "You know, Dr. Panofsky, load management is a new concept to us." The arrangement was accepted, but SLAC had to arrange metering at several points in the PG&E system in order to enable the company to judge when its total demand might threaten to become excessive.

The power line controversy interrupted what were almost uniformly excellent relations with the neighboring communities. In laying out the SLAC site, we established three different standards of architectural control. These, incidentally, had to be approved by Stanford's board of trustees because SLAC is on Stanford land. The central campus of SLAC had to meet the same standards as the on-campus buildings of Stanford University. Some of the buildings between the accelerator and the central SLAC campus meet what we call the "shop standard," which is comparable to normal industrial practice. The target area between the end stations and the backup hill is a "depressed area," meaning that it is excavated to a depth sufficient that is not visible to occupants of adjacent communities. Therefore, the activities related to physics research in the target areas need not meet specified architectural standards.

Early on, we had a complaint from a neighboring community that our modulators emitted annoying noises when in operation. We sent a SLAC staff member equipped with two-way radio communications to SLAC to the home of the complaining party. We found that the noise in question had no relation to the modulators being turned on or off, so the noise must have originated from some other, more mysterious source. SLAC received an award for being good neighbors from the Committee for Green Foothills, a neighborhood environmental protection association.

Construction proceeded, and thanks to Dick Neal's critical path management system and the good work of many SLAC staff members and contractors, a first beam through the entire two-mile machine was obtained on May 21, 1966. We attempted to increase both energy and current. The energy goal of 15–20 GeV specified in the proposal was soon exceeded,

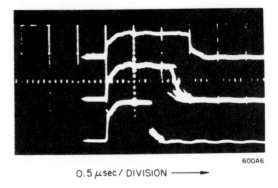

600A6

0.5 μsec / DIVISION ⟶

FIGURE 10.2. Beam breakup of the electron beam (or rather pulse-shortening) passing through the two-mile accelerator. Beam intensity is increased from the upper to
the lower oscilloscope trace. (From SLAC InfoMedia Solutions.)

but a difficulty arose when increasing the current. We encountered a phenomenon
we called "beam breakup;" this was exhibited through the beam disappearing
early during the pulse when the number of electrons per pulse exceeded a
certain amount. Figure 10.2 illustrates this phenomenon.

Beam breakup was not anticipated by any of the participants in the SLAC
design. It was known that a similar phenomenon occurred in single accelerator sections, but a cumulative re-generative beam breakup in a multisection
machine was not foreseen. I looked at this problem immediately in detail, and
developed a multicavity model that permitted a simplified analysis. In
essence, if the leading particles within the particle bunch are displaced off-
axis, they excite a mode in each cavity that leads to a further radial deflection
of the succeeding parts of the beam. The analytical solution was useful in
establishing the scaling laws for the onset of this phenomenon as a function
of beam intensity, accelerator length and gradient, and the strength of external focusing fields. The analytical approach to understanding this phenomenon[3] was followed by a numerical solution by Richard Helm. As a result of
this understanding, a number of cures for this phenomenon were devised that
were then designed in detail using Helm's numerical analyses.

Happily the problem was less severe than it might have been, since we had
chosen the "constant gradient" structure for each accelerating section. This
design employed a nonuniform structure designed to slow the filling of each
cavity as the energy of the traveling wave diminished. Such a nonuniform structure fortunately required some dispersion in frequency of the deflecting higher
modes. The simplest remedy was simply to strengthen the external focusing
fields supplied by magnetic quadrupoles. The second approach was to distort
some of the cavities in the accelerator such as to further disperse the frequencies of the higher radial modes but leave the frequency of the basic accelerating mode at its existing value. These remedies were applied rapidly. Further
remedies, applied later, were more accurate beam centering and further

strengthening of focusing. Therefore, beam breakup is no longer a limit to the intensity that can be generated by the SLAC two-mile accelerator.

The foregoing is a highly abbreviated account of some of the work that went into getting a beam through all the sections of the accelerator. Again, needless to say, this was an enormous effort of hundreds of people and I note that more than 90 authors contributed to "The Blue Book" describing the construction of SLAC. I now turn to the construction activities beyond the end of the accelerator.

The beam switchyard, which distributed beams to potential physicists-users, also was an enormous undertaking. It housed a forward beam and two magnetic achromatic bend systems called the A-beam and the B-beam. These had a similar mission to the achromatic translation systems discussed earlier, but terminated in beams at an angle to the primary beam, in contrast to a parallel translation. Again, an intermediate slit defined the energy and energy width of the beam to be transmitted to the experiments. Because it is expected that the pulse frequency acceptable to the various experiments will differ, the system permitted beam delivery interlaced in time achieved through a pulsed magnet designed to inject beams into each deflection system in accordance with a prearranged pattern. The transverse beam orbitry for the magnets was analyzed in detail by Karl L. Brown up to the second order. The actual construction of the magnet systems was a challenging enterprise and again, is described in detail in "The Blue Book."

A special problem was generated by the fact that the average power of the beam might be as high as 2 MW, and therefore, slits and collimators as well as the downstream beam dump had to be designed for high average power. As noted previously, Richard Taylor—who was in charge of the overall switchyard construction—made a number of conservative decisions in case activation near the beam intercept regions required handling of components by remote control. The beam dump consisted of a water-cooled set of copper plates coated with nickel and chromium. The beam-dump window was constructed to be remotely replaceable. Because of the high level of radioactivity expected even in the cooling water (including production of long-lived isotopes of beryllium), the cooling water was recirculated through a heat exchanger so that the primary cooling loop remained closed and shielded.

On October 2, 1964, the excavation of the B-beam in the beam switchyard was interrupted by a fortuitous event. One of our engineers, standing on the hill east of the target area, noticed that two bulldozers were passing one another as they were excavating for the B-beam housing; this passage should not have been possible according to the excavation width specified. The excavation was stopped for remeasuring, and at that time, fossil bones were seen exposed in the nearly vertical wall of the south embankment.

Earl L. Packard, a retired paleontology professor from Oregon State University who was visiting at the U.S. Geological Survey (USGS) branch at Menlo Park and at the Stanford geology department, was called in. He judged that the bones were of some large ancient animal and should be recovered.

Dick Neal called me in the East and we agreed that both excavation at the B-beam and collection of the bones could proceed because the trench had, in fact, been overexcavated, leaving room for both activities. Accordingly a "digging team" consisting of Packard, my wife Adele, a geology graduate student from Stanford, and Dr. Charles Repenning, a vertebrate paleontologist from the USGS, was assembled to collect the fossil.

SLAC construction proceeded without delay around the little shelter that was built for the paleontological collection, which took five weeks. Digging out the bones without damaging them is a delicate operation and some of them had to be embedded in a plastercast envelope to protect them.

A turn-of-the-century agreement between Stanford University and the University of California at Berkeley (UCB) provided that vertebrate specimens should be preserved at UCB in climate-controlled facilities. Accordingly, Stanford University then placed all their vertebrate fossils in the University of California Museum of Paleontology (UCMP) collections. The SLAC fossil was transferred to Berkeley, but in turn the Berkeley paleontologists made six excellent plaster casts, one set of which was presented to SLAC in exchange for the actual fossil, and another one to the USGS at Menlo Park; the others went to the Smithsonian Institution—National Museum of Natural History, and to the National Science Museum, Tokyo.

The paleontological find turned out to be a very important one. The specimen was identified as genus *Paleoparadoxia*, preserved from the Miocene period (about 14 million years ago) in the marine sandstone constituting the formation at SLAC east of the end of the accelerator. It was a unique find in the United States, both in terms of preservation and completeness of the skeleton. A smaller specimen existed in Japan. It was indeed "paradoxical" that *Paleoparadoxia* was the first scientific discovery at SLAC, preceding the results to come in physics.

A sentimental note: V. L. VanderHoof, the vertebrate paleontologist who had dedicated his mechanical skills to help physicists at UCRL during the war, and who became our good friend at Berkeley from 1945 on, had extensively studied the Miocene *Desmostylus*, an ancestor of *Paleoparadoxia*. So with this find at SLAC, physics reciprocated Van's earlier assistance to physics by contributing to paleontology along the lines of his research interests. Sadly, Van had died of lung cancer a few months before the SLAC find.

The plaster cast of *Paleoparadoxia* was prepared for exhibit at SLAC's Visitor Center by my self-educated paleontologist wife, Adele. It took her over 20 years, interrupted by such details as family raising and taking care of a frequently itinerant husband. Her work is documented in detail in a SLAC publication.[4]

I did worry that if paleontologists behaved like high-energy physicists, they would swarm to SLAC, demanding opportunities to dig. That did not happen, so the construction schedule remained unperturbed by the discovery of *Paleoparadoxia*.

Not only was the beam switchyard enclosure a very massive earth-covered concrete structure, but the buildings to house the experiments downstream also presented engineering challenges. Those buildings were designed in parallel with the design and construction of the experimental equipment that they were to house. In turn, each piece of experimental equipment was planned, generally by a SLAC faculty member, for a specific experimental program, as discussed with the program advisory committee.

The two target areas, known as End Station A and End Station B, were designed to house experiments, one using the primary electron or photon beam (End Station A), and the other utilizing secondary particles produced by the beam (End Station B). I was personally concerned in the design of End Station A and interacted extensively with the group headed by Richard E. Taylor in defining the basic design concepts of the spectrometers to be housed in that building. Architecturally, End Station A was a very difficult problem because we were not sure where and how various beams might have to be transported beyond the walls of the building. At the same time, the building required heavy shielding to confine the radiation if a significant fraction of the primary beam were to be stopped in the building. Accordingly, the walls of the building were designed to be removable, requiring that the massive load of the thick roof—needed for containing radiation—be carried entirely by the corner posts of the building (because the walls were removable). We also avoided any columns inside the building, again because the experimental arrangements needed to remain flexible. Therefore, the ceiling load had to be carried by very long, prestressed, upward-curved concrete beams. Interestingly enough, these engineering problems were solved and the resulting massive building received an award for excellence in architecture!

End Station B was smaller, but of similar construction to that of End Station A, and was designed primarily for the bubble chamber program. As was discussed previously, we undertook the design and eventual construction of the major instruments needed for particle research in parallel with building the accelerator, the switchyard, and the research areas.

Three particle spectrometers were built in End Station A to study electron scattering and photoproduction. In contrast to the electron scattering spectrometer at HEPL, all three SLAC spectrometers incorporated line-to-point focusing, because the anticipated targets were of low density, presumably liquid hydrogen or deuterium, and would therefore be elongated. Each of the three spectrometers was designed for a progression of energies of the particles to be studied, and their angular range was tailored to fit the anticipated kinematics. Note that this arrangement differed sharply from that envisaged in the proposal, which showed one large spectrometer covering the entire angular range from 0 to 180 degrees. The smallest of the spectrometers was designed by a research group headed by David Ritson from the Stanford physics department, in cooperation with physicists from Northeastern University. The other two were designed and built by a collaboration among SLAC, MIT, and Caltech, headed by Richard E. Taylor.

The three spectrometers rotated around the same central pivot, which was constructed using parts of a surplus large naval gun. This arrangement served a large number of programs well, but has now been superseded by experimental arrangements serving other specific purposes.

Let me make some comments on hydrogen bubble chambers at SLAC. Joseph Ballam designed a one-meter rapid-cycling bubble chamber to examine secondary particle interactions, and he designed a monochromatic gamma-ray beam to initiate production of such particles. That beam was formed by backscattering photons from an intense laser beam from the primary electron beam. At the time, Luis Alvarez in Berkeley was operating the much larger (72-inch) hydrogen bubble chamber in an enormously successful program at the Berkeley Radiation Laboratory. Alvarez received the 1968 Nobel Prize for this productive work. In fact, the program was so successful that in the mid-1950s Alvarez tried to persuade the then-director, Edwin McMillan, to cancel all other means of detecting high-energy particles at Berkeley because the liquid hydrogen bubble chamber would at any rate wipe them off the map. McMillan was not persuaded by that argument, and he arranged for a public debate between Luis Alvarez and me which drew quite a large audience. Luis asked me whether I could think of any experiments using the Berkeley accelerators that could not be done best by his bubble chambers. I indeed recited some, such as scattering processes involving large momentum transfer or the production of secondary particles at a very small branching ratio. Alvarez' reply was that these were relatively minor exceptions, and after all (a flattery!) it took WKHP to think of them. McMillan's response was that Alvarez should "turn his collar backwards."[5] In consequence, several other means of detection continued productive work at Berkeley.

In subsequent conversations with Joe Ballam, Luis Alvarez concluded that the 72-inch bubble chamber could greatly increase its effectiveness and data rate if it were transferred to SLAC. At Berkeley, the repetition rate of taking pictures was controlled by the pulse rate of the Bevatron; whereas at SLAC, the picture rate would be controlled by the expansion rate of the bubble chamber, because the beam repetition rate at SLAC was many times larger than the rate the chamber could accept. This idea led to a program improving the 72-inch chamber to a somewhat higher pulse rate and lengthening it to 82 inches. The chamber then, figuratively speaking, walked across the Bay in 1967, and became an extremely productive tool at SLAC, continuing its past success at Berkeley. Several physicists from Berkeley also "walked across the Bay" with the bubble chamber.

The 82-inch bubble chamber was accompanied by Robert (Bob) Watt, a senior technician who had learned cryogenic and bubble chamber practice at Berkeley while serving there as Luis' chief operator. Before that period, he worked as chief operator of the 32-foot proton linear accelerator when I was still at UCRL. Bob was a great asset to SLAC. He understood physics and engineering relevant to the chambers better than most professionals and he was very safety conscious. I recall when a safety inspection of the bubble chamber complex was conducted by an AEC team. One of the inspectors complained about the lack of a water sprinkling system over the hydrogen chambers and the associated cryogenic installations. Bob laconically replied, illustrating his remarks with directional gestures: "Hydrogen up ↑; water down ↓." The safety team had to agree. Bob worked for SLAC throughout the period the chambers stayed in operation and worked in other cryogenic activities thereafter.

Liquid hydrogen was delivered to SLAC in large tankers. On one occasion, a supply pipe on one of the trucks broke after arriving at SLAC and a flame erupted at the fracture. I was called at home and drove to SLAC only to find all roads blocked by police and fire personnel; an alarm had been issued largely based on the frightening word "hydrogen," warning of a possible "H" explosion and invoking images of the "H" bomb. I talked my way in and suggested moving the truck to an area remote from any buildings. But then Bob Watt appeared with a helium cylinder: he hooked the helium supply to the truck's piping, the helium displaced the hydrogen, and the flame went out. End of panic.

At Berkeley, the maximum bubble chamber picture production rate had been about one and half million pictures per year; this rate was quadrupled at SLAC. This increase resulted in a period of several years where the data analysis facilities worldwide were largely saturated with the analysis of SLAC pictures. The development of the film generated by the bubble chambers at SLAC was still carried out at Berkeley, and one of the SLAC guards was charged with routinely driving the exposed film from SLAC to the Lawrence Berkeley Laboratory to be developed. When the guard returned from one of his trips, I asked him whether he had any trouble, and he said, "Oh no, but I had to draw my gun to get inside because there were so many demonstrators." This scared me, and I decided to disarm SLAC by forbidding any firearms on the site. I had hoped that this would be consonant with university policy, but found out to my surprise that the university was unwilling to proclaim a similar ban on the main campus because some of the resident faculty members wanted to keep their guns.

Another large installation planned and built concurrently with SLAC construction was the two-meter streamer chamber built by Robert Mozley. It was used in the central beam. Another large spectrometer was a multiplate spark chamber arrangement built under the direction of David Leith which also served a large multitude of experiments.

To summarize, the construction of the SLAC accelerator was paralleled by building a formidable array of instruments, which in turn could support a large community of physicists both inside and outside the laboratory. The successful work in accelerator construction was well recognized outside the laboratory even before the research program started. I received the National Medal of Science in 1969, both in recognition of the completion of SLAC (in particular without breaking the budget) but also for arms control work and science advising. President Nixon gave me the award, and he acknowledged Adele's red suit by commenting that she was wearing "Stanford red."

I describe the research work carried out with those instruments and others in a later chapter, but now return to a recital of activities external to SLAC that paralleled SLAC's construction and initial operation.

11
Physics and the Cold War

As construction of SLAC proceeded,[1] outside commitments did not diminish. My PSAC membership did not terminate until 1964, and I continued to serve as a consultant to the White House's Office of Science and Technology until 1973. In parallel, in 1965 I was recruited to serve as a member of JASON. As is well known, the JASON group is composed of academics who, particularly during the summer months, undertake independent studies on national security matters for various governmental bodies

During this same time period, communication and collaboration in high-energy particle physics itself became progressively international, and SLAC became an increasingly prominent part of the worldwide program. In 1956, High Energy Physics (HEP) even pioneered in a friendly penetration of the Iron Curtain. A group of 14 U.S. physicists, including me, were invited to visit the HEP laboratories in the Soviet Union. Our first scheduled visit was in Moscow, but our Scandinavian Air Lines flight had to make an emergency landing in Riga en route. While the plane was waiting for the arrival of a replacement part, we stayed at the airport, where the local personnel were very confused about how to treat us, but all went well. When we finally arrived in Moscow, we toured a research reactor, the facilities at the Institute for Theoretical and Experimental Physics (ITEP) and then Gersh Budker's Research Institute, which at the time was still in Moscow. We were introduced to senior physicists who had also been major figures in the Soviet nuclear weapons program. I remember a conversation I had with Lev Artsimovitch; we discussed the not-so-serious proposal by Enrico Fermi to build an accelerator in the vacuum of outer space encircling the earth. Artsimovitch asked whether I had estimated the cost of such a device. I replied, "The sum of the U.S. and the Soviet military budgets would pay for it in a few years." He changed the subject.

We then were driven to Dubna, the international laboratory for the "Socialist Countries." A car with a physician followed, just in case one of the visitors became ill. As we walked down some of the streets, the local inhabitants were kept behind barricades, and viewed us as if we were visitors from other planets. But we were well and cordially received and given tours of the so-called "synchrophasotron," a synchrocyclotron in U.S. terminology, and the then-incomplete 10-GeV proton synchrotron.

I was impressed by the profusion of beam lines, the instrumentation, and the generous proportions and fabrication practices of the installation. But there was also a lack of originality in the experimental programs: we saw no experimental setups at either ITEP or Dubna that were not patterned after prior Western work.

An exception to this observation was Budker's laboratory in Moscow, which was undertaking pioneering efforts in some daring new technologies: plasma confinement experiments, pulsed magnets with boundaries defined by eddy-current sheets, and other novel techniques. Here the endeavors were quite opposite to those of the other laboratories; some quite daring but possibly not immediately applicable technologies were being pursued, rather than the very conservative instrumentation seen at the other laboratories.

The cordiality and personal warmth of our reception in the Soviet Union was excellent and clearly sincere. We were hosted at some of the scientists' homes. I recall a visit to Professor Venedikt Dzhelepov, the physicist responsible for building the synchrophasotron, where I noticed a bear rug hung on the wall. On inquiring as to its origin, I received the answer, "My grandmother shot him while she was working on the railroad." Dzhelepov reciprocated that visit at our home in California several years later.

This visit began a new era of communications in high-energy physics. Vladimir Veksler, the Russian co-inventor of phase stability, was invited to the next Rochester Conference in the United States, and there he gave a memorable talk in which he remarked, "There are now three branches of physics: experimental physics, theoretical physics, and diplomatic physics."

Cooperation in high-energy particle physics became elevated to a high level. A Soviet delegation headed by Andronik M. Petros'yants, chairman of the Soviet State Committee on the Utilization of Atomic Energy, visited SLAC in late November 1963. He was accompanied by a number of aides and the United States was represented by John Teem, the director of research of the AEC.

I accompanied Petros'yants on a private visit to San Francisco. He confided in me that he was instructed by his wife to procure baby bottle nipples for his grandchild which were in short supply in the Soviet Union, and to the extent available, didn't work well. In response, my wife and I took him to an all-night grocery store in San Francisco. Because separate nipples were not available, he purchased a large bag of complete American baby bottles and seemed greatly pleased by having fulfilled his mission.

Petros'yants visited SLAC together with his entourage and the accompanying American group late in November 1963 just when President Kennedy was assassinated. The U.S. government was greatly concerned about his personal safety in the wake of the assassination because there was fear that some U.S. citizens might lash out against any Soviet visitors then in the United States. Accordingly, we spirited Petros'yants off to Yosemite Valley to keep him in hiding during the turbulent days after the assassination. Nothing adverse happened and Petros'yants returned safely to his homeland.

Another outside activity during this era was my participation in the antiballistic missile defense debates of 1968. Since the beginning of the nuclear age, the issue of offense versus defense kept resurfacing. The demands on any defense against nuclear weapons are extremely difficult to satisfy if the defense is to be reasonably effective in relation to its cost; yet the pressures for defenses against nuclear weapons delivered by intercontinental ballistic missiles continue until this day. Unfortunately, the debates about antinuclear defenses continue to be extremely politicized. Somehow, no administration or Congress can resist the argument, "How can we possibly leave this nation undefended against nuclear attack?" even if rational consideration indicates that the specific defenses under discussion are most unpromising.

Defenses are in two categories: passive measures to mitigate the effects of nuclear weapons should a nuclear attack occur; and active attempts to interdict the delivery of nuclear weapons onto United States' or Allied soils. In 1956, proposals were made to initiate an extensive civil defense shelter program in the United States. Such programs can range from providing shielding against radioactive fallout far from nuclear explosions to shelters offering some protection from blast and fire in regions more proximate to the detonation. In response to these proposals, President Eisenhower established the Gaither Committee to review the utility of the enormously expensive and extensive proposed shelter deployment. The Gaither Committee did so, but also went beyond its charge and looked at the whole gamut of issues affecting the vulnerability of the United States to nuclear attack. The Gaither Report painted a grim picture.

The offense–defense balance was further addressed in 1959 when the President's Science Advisory Committee convened a panel, of which I was a member, on antiballistic defense. At that time, the military services had proposed an antiballistic missile system called Nike–Zeus to defend 27 areas in the United States with 7000 deployed interceptor missiles carrying nuclear warheads. In its classified report, the panel, chaired by Jerome Wiesner, identified the weaknesses of the Nike–Zeus system; these included the vulnerabilities of the radars controlling the interceptor missiles as well as a series of countermeasures that the offense could deploy.

The issue is basic: even a single nuclear explosion on United States soil detonated in or near an urban area can kill up to a million people, and maim many more. Moreover, the attacker can choose where to attack, and therefore, to offer any real protection, a defense has to be very inclusive. Moreover, once the character of a defense is known to a potential attacker, he can generally adopt either alternative countermeasures to defeat the known defense or a means to bypass the defense altogether. Therefore, what is now called the "defense effectiveness on the margin" has—for all means of defense against nuclear weapons—been unfavorable to the defense: once a defense has been deployed, the opponent can augment or modify the offensive measures, leaving the defended nation just as vulnerable as before, but this change in offense is very much less expensive than the cost of the defense.

The PSAC Panel identified all these weaknesses of Nike–Zeus, and these lessons were directly communicated to President Kennedy when Jerome Wiesner became his science advisor in 1961. By that time, the military services proposed major changes in the Nike–Zeus program by adding a so-called "phased-array radar," which could direct a large number of interceptors, including both long-range and short-range missiles. The President was persuaded that this system also had major problems, yet he supported the program with generous research and development funds, while withholding any deployment decision. Kennedy's Secretary of Defense, Robert McNamara, understood these issues well, and convinced Kennedy to consider only a reduced Nike–Zeus system consisting of perhaps 1200 intercept missiles. It was therefore clear to all parties that such a system would not be adequate to interdict an all-out attack from the Soviet Union.

But then the military developed the concept of a "thin" defense, called "Sentinel," which might defend the United States against a very limited attack from what was at that time the "rogue" nation, namely communist China. At the same time, Kennedy and McNamara became convinced that the only way to limit or even to reverse the expensive and dangerous nuclear arms race between the United States and the Soviet Union was arms control. McNamara met with Alexei Kosygin, the prime minister of the Soviet Union, in Glassboro, New Jersey, and tried to persuade him that limiting defenses by mutual agreement would be a constructive way to limit the arms race. Kosygin was unswayed, and he is quoted as having said, "When I have trouble sleeping, it's because of your offensive missiles, not your defensive missiles."

In September 1967, McNamara made a speech in San Francisco where he outlined the many deficiencies of ballistic missile defenses, but then to everyone's surprise, he ended that speech by endorsing the deployment of the "thin" Sentinel defense to counter the China threat. At the same time, he held up no hope that a Soviet missile attack could be blunted. McNamara's speech ignited a national debate in 1968, and for the first time, senior United States senators sought the advice of individual independent scientists rather than relying solely on briefings from administration witnesses. As a result, I and several other physicists, most of them current or former members of PSAC, had numerous and extended conversations with Albert Gore, Sr., the senator from Tennessee, and the very hawkish but very realistic Stuart Symington, the senator from Missouri.

Then the administration changed in 1969: Nixon became president, and David Packard became deputy secretary of defense. I had known Packard fairly well from his former roles as both chairman of the board of trustees of Stanford University, and as a prominent engineer in Silicon Valley. On a personal note, as he prepared to leave for Washington, Packard called me into his office and persuaded me to become co-chairman, with Edward Ginzton, of the local branch of the Urban Coalition; this was an organization dedicated to ameliorating the racial problems on the San Francisco Peninsula, including the poverty prevailing in East Palo Alto. I worked diligently with Ginzton on this assignment for two years and we made significant progress, in particular in respect to strengthening the educational opportunities in East Palo Alto.

After Packard assumed his office in Washington, the Defense Department changed the mission of Sentinel to what then became known as "Safeguard," a system designed to defend the silos housing intercontinental ballistic missiles (ICBMs) against a possible first strike attack by Soviet missiles. But the problem now became that the defense department had changed the mission but not the hardware to accomplish that mission. I personally considered the defense of ICBM silos the only application of missile defense that might be justified. Such defenses—to preserve the United States' deterrent forces in case of a Soviet disarming attack—would in principle be possible. However, even a successful attack against all ICBM silos would still preserve the airborne and submarine components of the American strategic "triad." Therefore, if you consider the time sequence required for attacking the different "legs" of the triad, such an attack against the land-based missiles, even if it could succeed, would be totally foolhardy and suicidal on the part of the Soviets.

As it happened, I ran into David Packard at the San Francisco Ambassador Club, the TWA frequent flyer lounge, as we both were headed out of town on a trip to Washington. He asked me what I thought about the change of mission from Sentinel to Safeguard. I replied that the change of mission was indeed an improvement, but that the hardware which he had inherited was totally unsuitable for that mission, and we discussed these issues further.

During a subsequent visit to Washington which I had undertaken to talk to members of the Joint Committee on Atomic Energy on SLAC-related matters, I had some spare time and attended a hearing where Packard was testifying before the Senate Committee on Foreign Relations, chaired by Senator Gore, Sr. Packard explained the Safeguard program, and Gore asked, "From whom did you get your scientific advice?" Packard said, "From Professor Panofsky," and Gore asked where, and he said, "At the Ambassador's Club." This exchange caused a great deal of hilarity and led to a column in the *San Francisco Chronicle* by Arthur Hoppe which joked that the method by which the Pentagon got advice on sensitive military matters was by trapping unwary scientific travelers at airports.

But to return to the hearing: one of Gore's aides reminded the senator that I was in the audience, and Gore asked me to testify the next day. After thus having been "Gored," I spent the night preparing testimony in which I tried to explain to the committee that the hardware was unsuitable to the stated mission. The phased-array radars could be blinded by a single precursor nuclear explosion, and in addition, the number of interceptor missiles at any one U.S. ICBM site was totally inadequate and could be exhausted by a relatively small number of Soviet warheads or decoys.

This testimony was given against a background of major controversies on nuclear weapons policy within the U.S. government. A group of analysts, of whom one of the most prominent was Albert Wohlstetter, was engaging in very detailed numerical analyses, attempting to fine-tune the conduct and consequences of protracted nuclear war, and defenses continued to play a substantial role in these detailed analyses. One of the individuals testifying before Congress as this same time was George Rathjens, a member of the staff of PSAC who later became the secretary general of The Pugwash Conferences on Science and World Affairs. Rathjens gave testimony on his views on the vulnerability of the U.S. ICBM silos based on assumptions about their blast resistance and the number of warheads available to the Soviet Union to attack the silos.

Wohlstetter disagreed both with my testimony as well as that of Rathjens and others, and he filed an official complaint with the Operations Research Society of American (ORSA) that Rathjens and other prominent scientists, including me, were "practicing operations research without a license." The complaint was mainly directed against Rathjens, who unfortunately had used figures somewhat different from those accepted for the vulnerability of missile silos to blasts and for the number of available warheads. To me, these detailed numbers appeared largely irrelevant for evaluating the deficiencies of ABM, but ORSA took these differences between Rathjens and Wohlstetter in particular most seriously.

Accordingly, the board of ORSA adopted a resolution censoring the scientists' testimony. I considered this matter to be somewhat of a joke, inasmuch as I was not an ORSA member, and at any rate, operations research had been initiated during World War II by physicists who introduced analytical methods into the evaluation of military operations. Philip Morse, a founding member of ORSA, resigned in protest over the ORSA board's action.

The rest is history. Congress and the administration approved the limited Safeguard deployments, and in 1972 the United States and the Soviet Union signed the ABM Treaty, which limited ABM deployment to 200 interceptors on each side. This was later reduced to 100 interceptors by a 1974 amendment. The Soviet Union had deployed a limited defense incorporating 64 interceptors around Moscow. That deployment still exists, although it is believed to be in bad shape; however, Safeguard was cancelled as being not worth the money after being operational for less than a year.

This story continues to be replayed in subsequent events and in a variety of contexts. It is part of the general divergence in view as to what role, if any, nuclear weapons, and defenses against them, should play within the national nuclear strategy. On one end of the spectrum of opinion at the time of the ABM debate were those who concluded that nuclear weapons could not actually be used in warfare, but that their only role was to deter the use of nuclear weapons by making sure that there was broad recognition of their enormous destructive power and the extreme vulnerability of countries to the exercise of that power. In popular terms, this situation was described as "Mutual Assured Destruction" or "MAD."

On the other end of the spectrum of opinion were the large groups of studies attempting to fine-tune the operational use of nuclear weapons in a war where their use could be either intermingled with or substituted for conventional weapons. I discussed this tension in an article[2] written together with Spurgeon M. Keeny, Jr., entitled "MAD vs. NUTS" where "NUTS" stands for "Nuclear Utilization Target Selection." Spurgeon and I had maintained a continuing collaboration following his role supporting PSAC and the Geneva negotiations. That article, as well as many other future communications, attempted to make clear that the destructive effects of even a small number of nuclear weapons, including collateral effects such as fire, fallout, and societal disruption reaching much beyond the intended target, were so large that the gap between MAD and NUTS was much narrower than many military analysts, who were not sensitive to the realities of nuclear weapons, tended to assume.

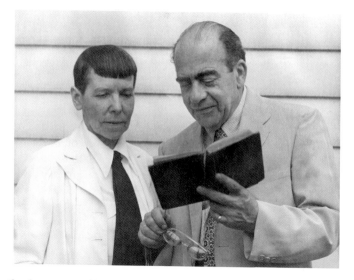

The author's parents Dora and Erwin Panofsky. (Credit: Panofsky Family Collection.)

The author in Hamburg at age about 6 months. (Credit: Panofsky Family Collection.)

The author at age about 4 years playing chess with his cousin Ruth Mosse, while his brother Hans looks on. (Credit: Panofsky Family Collection.)

The author at four years old playing chess. (Credit: Panofsky Family Collection.)

Class photo at Fraülein Lehrmann's Vorschule about 1924. The author is the fourth pupil from the left in the front row. (Credit: Panofsky Family Collection.)

Jesse W. M. Du Mond, the author's Ph.D. supervisor and father-in-law. (Credit: Harvey of Pasadena photo, Panofsky Family Collection.)

The author and fellow Caltech graduate students: (left to right) Vernon Hughes, WKHP, William Eberhard, Edward Deeds, at a Southern California mountain top, circa 1940. (Credit: SLAC Archives and History Office, Panofsky Collection.)

The author and fellow graduate student Donald Wheeler playing chess during a long finishing cut on the lathe. Caltech, circa 1940. (Credit: SLAC Archives and History Office, Panofsky Collection.)

Wedding photo in Du Mond Garden, 1942. (Credit: Panofsky Family Collection.)

Adele with the twins, Richard and Margaret, 1943. (Credit: Panofsky Family Collection.)

The team building the 32-MEV proton linac at the University of California Radiation Laboratory sitting atop the 40-foot vacuum tube housing the accelerator. The author is seated as #17 from the left, next to Robert Serber. Louis Alvarez is #25 from left. (Credit: Lawrence Berkeley National Laboratory.)

Luis Alvarez and the author holding a coupling loop transferring power to a 200-Megahertz resonant cavity. (Credit: Lawrence Berkeley National Laboratory.)

Meeting of Technical Working Group II of Comprehensive Test Ban Treaty at United Nations HQ, Geneva, 1954. Jim Fisk, Chair, flanked by Vice-Chair WKHP on left, and Doyle Northrup on right. Harold Brown on left of WKHP. John Tukey and Hans Bethe behind Fisk. (Credit: SLAC Archives and History Office, Panofsky Collection.)

Group photograph of first Western visit to Soviet Laboratories in 1956. Left to right Owen Chamberlain, Vladimir Veksler, Mark Oliphant, Luis Alvarez, Venedik Dzhelepov, five unidentified, WKHP, Jack Steinberger, Isaak Pomeranchuk. (Credit Budker Institute of Nuclear Physics photo, SLAC Archives and History Office Panofsky Collection.)

Issak Pomeranchuk facing Rudolph Peierls with Arkadii Benediktovish Migda (center) and Lev Landau (right) looking on, 1956 meeting of Soviet and American scientists. (Credit: Budker Institute of Nuclear Physics photo, SLAC Archives and History Office, Panofsky Collection.)

Vitali Goldanski, Emilio Segré, Pavel Cerenkov, two unknown, and WKHP at 1956 meeting. (Credit: Budker Institute of Nuclear Physics photo, SLAC Archives and History Office, Panofsky Collection.)

Left to right: Pavel Cerenkov and his camera, two unknown, WKHP, and Owen Chamberlain. (Credit: Budker Institute of Nuclear Physics photo, SLAC Archives and History Office, Panofsky Collection.)

The five Panofsky children at Lyell Fork in the High Sierra, 1957. (Credit: Panofsky Family Collection.)

The Panofsky children in monogrammed sweaters prior to trip into the Stehiken wilderness, 1958. (Credit: Panofsky Family Collection.)

Signing of Technical Working Group II agreement, 1959. Fisk flanked by WKHP and Doyle Northrup. Frank Press, Anthony Turkevich, Hans Bethe, and John Tukey are behind. (Credit: SLAC Archives and History Office, Panofsky Collection.)

A meeting of the President's Science Advisory Committee in Newport News. July 12 1960. WKHP presenting nuclear test ban report to President Eisenhower. Clockwise from left to right: Mannie Piore, Don Hornig, George Kistiakowski, President Eisenhower, George Beadle, unknown, John Tukey, unknown, John Bardeen Jim Killian, Al Weinberg, WKHP, Jerry Wiesner, Wally Z, Detlev Bronk. (Credit SLAC Archives and History Office, Panofsky Collection.)

Panofsky children on tigers in front of Nasson Hall, Princeton University Campus, August 1962. (Credit: Panofsky Family Collection.)

From left: Marvin Chodorow, WKHP, Robert Minge Brown (Counsel to Stanford Board of Trustees), Ed Ginzton, 1962. (Credit: Richard Muffley photo, SLAC Archives and History Office.)

Felix Bloch, and WKHP at picnic on SLAC site, 1962. (Credit: SLAC Archives and History Office, Panofsky Collection.)

Pief and Adele on SLAC site, 1962. (Credit: Richard Muffley photo, SLAC Archives and History Office.)

Pief, Adele, and Hobey De Staebler at SLAC picnic, 1962. (Credit: Richard Muffley photo, SLAC Archives and History Office.)

Marvin Chodorow, Larry Mohr (AEC), and WKHP at SLAC picnic, 1962. (Credit: Richard Muffley photo, SLAC Archives and History Office.)

Morris Doyle, Chairman of Stanford Board of Trustees, signing SLAC construction contract. Dwight Adams (Stanford V.P. for Business) and WKHP look on. Robert Minge Brown (Stanford Counsel) shows contract to Ira Lillick (Board Secretary). April 30, 1962. (Credit: Stanford University News Service.)

Meeting of President's Science Advisory Committee (PSAC) at Executive Office Building next to White House, Washington D.C., 1963. Jerome Wiesner, Presiding. (Credit: Cecil W. Stoughton photo, SLAC Archives and History Office, Panofsky Collection.)

US-National Academy of Sciences delegation at bilateral meeting at the Soviet Academy in Moscow to negotiate scientific cooperation between the U.S. and the U.S.S.R., 1967. From left to right, WKHP, George Kistiakowski, Paul Doty, two unknown. Harrison Brown. (Credit: SLAC Archives and History Office, Panofsky Collection.)

SLAC Dedication, 1967. Glenn Seaborg (Chair, AEC), WKHP, Wallace Sterling, Don Hornig (President's Science Advisor), Ed Ginzton. (Credit: Stanford University News Service.)

From left: Congressman Chet Holifield, Congressman Craig Hosmer, WKHP, Ed Ginzton at SLAC Dedication, September 1967. (Credit: Stanford University News Service.)

Venedikt Dzhelepov at SLAC Dedication, 1967. (Credit: SLAC Archives and History Office, Panofsky Collection.)

Adele on Fire Engine 98 at SLAC Open House, September 1967. (Credit: Stanford University News Service.)

Old hand-pumper fire engine on exhibit as spoof on the governmental cost-reduction program. The Stanford fire chief and crew on left; FVL Pindar, Larry Mohr (AEC Area Manager), and WKHP at right, 1967. (Credit: Stanford University News Service.)

Matt Sands (SLAC Deputy Director) and WKHP, 1969. (Credit: Panofsky Family Collection.)

The 82-inch liquid hydrogen bubble chamber after transfer from U.C. Berkeley to SLAC, 1970. On the steps, left to right: Luis Alvarez, Robert Watt, Joseph Ballam, and WKHP. (Credit: Walter Zawojski photo, SLAC Archives and History Office.)

Leon Lederman, WKHP, Provost Terman, and Al Silverman during SLAC SPC
meeting, circa 1970. (Credit: Panofsky Family Collection.)

WKHP with Melvin Schwartz and I. I. Rabi at Columbia University, date unknown. (Credit: Panofsky Family Collection.)

WKHP with I. I. Rabi, Columbia University, date unknown. (Credit: Panofsky Family Collection.)

Soviet Delegation visiting SLAC in 1972, led by A. M. Petros'yants, Chairman of U.S.S.R. State Committee for the Utilization of Atomic Energy. Left to Right: Sid Drell, Al Lisin, Dick Neal, unknown, WKHP, John Teem (Director of Energy Research, AEC), Petros'yants, Richard Taylor, Larry Mohr (AEC). (Credit: Walter Zawojski photo, SLAC Archives and History Office.)

Zhang Wenyu visited SLAC in November 1972 to explore the best approach for China to enter the world of accelerator-based HEP. (Credit: Walter Zawojski photo, SLAC Archives and History Office.)

Zhang Wenyu and Chinese Delegation at SLAC, 1972. (Credit: Walter Zawojski photo, SLAC Archives and History Office.)

WKHP and Ed McMillan at APS meeting at the Fairmont Hotel in San Francisco in 1973. (Credit: SLAC Archives and History Office, Panofsky Collection.)

WKHP with Gersh Budker at meeting in Novosibirsk, 1975. A painting of Kurchatov looms overhead. (Credit: SLAC Archives and History Office, Panofsky Collection.)

Dedication of new computer building at SLAC. Provost Bill Miller and AEC Research Director Teem on WKHP's right, 1975. (Credit: Walter Zawojski photo, SLAC Archives and History Office.)

Groundbreaking for the Position–Electron Project (PEP), June 1977. Senator Alan Cranston wields the shovel. John Rees behind WKHP on his right. (Credit: Walter Zawojski photo, SLAC Archives and History Office.)

WKHP at lectern, PEP dedication, June 1977. (Credit: SLAC Archives and History Office, Panofsky Collection.)

Left to right: Unknown, Paul Doty, Pyotr Kapitza and Harrison Brown at Kapitza's Dacha, about 1980. (Credit: Panofsky Family Collection.)

Visit with Chinese Science Minister Fang Yi at the Great Hall of the People, October 14, 1983. Adele at WKHP's right. (Credit: Panofsky Family Collection.)

With Benyamin Siderov and Sasha Skrinsky in Novosibirsk, 1985. (Credit: Panofsky Family Collection.)

With Burt Richter, Ben Siderov, and Nikolai Dykansky in Novosibirsk, 1985. (Credit: Panofsky Family Collection.)

WKHP, Nikolai Dikansky, and Burt Richter in Novosibirsk, 1985. (Credit: Panofsky Family Collection.)

Tetsuji Nishikawa and WKHP in Novosibirsk, 1985. (Credit: Panofsky Family Collection.)

WKHP with Hans Bethe at Galvez House library, Stanford University, late 1980's. (Credit: Panofsky Family Collection.)

NAS visit to China, 1988. Left to right: General Pan, Mme Zhou Yunhua, unknown, Zhu Guangya, WKHP, T.D. Lee, Du Xianwan, Jeanette Li. (Credit: Panofsky Family Collection.)

Sid Drell Andrei Sakharov, and WKHP at Stanford University, September 2, 1989. (Credit: Harvey Lynch.)

Bill Wallenmeyer, Jim Leiss, and WKHP, about 1990. (Credit: Panofsky Family Collection.)

SLAC Chief Engineer Robert Gould and his cartoons of SLAC construction with the Wizard Gandalf the Grey, 1990. (Credit: Tom Nakashima photo, SLAC Archives and History Office.)

WKHP at 1990 Nobel ceremony in Stockholm with stuffed emperor penguin of same size. (Computer generated). (Credit: SLAC Archives and History Office, Panofsky Collection.)

Unknown, WKHP, unknown, and Academician Hu Side (Chinese Academy of Engineering Physics) at Lawrence Livermore National Lab, 1994. (Credit: Panofsky Family Collection.)

WKHP at the Chinese Institute of High Energy Physics in Beijing, March 1999. (Credit: Panofsky Family Collection.)

Amaldi Meeting on Problems of Global Security in Mainz, Germany, October 1999. From left to right: WKHP, Goetz Neuneck, and Jo Husbands facing Al Narath. (Credit: Panofsky Family Collection.)

Former Director of the Chinese Institute of High Energy Physics Zhen Jipeng and WKHP in Beijing, March, 2000. (Credit: Panofsky Family Collection.)

WKHP and Xie Jialin in Beijing, March 2000. (Credit: Panofsky Family Collection.)

WKHP at DESY, January 2003. (Credit: Panofsky Family Collection.)

WKHP and Francesco Calogero at Amaldi Meeting, Helsinki, 2003. (Credit: Panofsky Family Collection.)

With T. D. Lee and Peter Rosen at the 25th Anniversary US-PRC collaboration meeting in Beijing, October 2004. (Credit: Fred Harris.)

WKHP among giants: Tom Kirk and Chen Hesheng, 2004. (Credit: Fred Harris.)

Adele Panofsky on "Tar-baby," North Rim, Grand Canyon, August 2004. (Credit: Panofsky Family Collection.)

The author and future wife Adele on the steps of the Athenaeum at Caltech on th
day of the Ph.D. ceremony, 1942. (Credit: Panofsky Family Collection.)

WKHP and Adele on the steps of Caltech's Athenaeum in May 1999, 57 years afte
WKHP's graduation photo at the same place. (Credit: Panofsky Family Collection.)

Watercolor by Betty Martinelli, wife of the author's graduate student Emest Martinelli, depicting the move from U.C. Berkeley to Stanford in 1951 in response to the Loyalty Oath. The vehicle is an ancient V-12 Cadillac towing a Jeep. (Credit: Panofsky Family Collection.)

WKH with Pope John-Paul II, March 1999, on the occasion of the Papal Academy Meeting, headed by Nickolai Cabbibo (next to Pope). (Credit: photo by *L'Osservatore Romano*.)

The succession of SLAC Directors: Jonathan Dorfan, Burton Richter, and WKHP. September 1999. (Credit: John Ashton photo, Panofsky Family Collection.)

Adele with Paleoparadoxia model at SLAC Visitor Center, 1996. (Credit: Panofsky Family Collection.)

WKHP and Adele in California Desert with crested barrel cactus, 2005. (Credit: Panofsky Family Collection

12
Student Unrest at Stanford

Subsequent to the completion of SLAC and before the research program had acquired much momentum, a period of student unrest had a large impact on Stanford University, including SLAC. I was drawn into these conflicts in a number of ways. We adopted a liberal policy in letting protesters conduct meetings in the SLAC auditorium. Some of the resulting discussions, which were—amazingly enough—conducted in a civilized manner, turned out to be constructive and useful in relaxing some of the tensions. Nevertheless, I was called to account by members of the Congressional Joint Committee on Atomic Energy asking me to explain why government property, meaning the SLAC auditorium, was used by the student agitators. I explained the situation as being consistent with university policy and as being constructive in purpose; I'm happy to say that particularly Congressman Holifield and others were satisfied with that explanation.

While I was in Washington eating lunch in the AEC cafeteria on December 7, 1971, I overheard a conversation at a neighboring table, "Have you heard about the explosion at SLAC?" Obviously I chased this down, and found that indeed during the early morning hours an explosive device had damaged the first section of the equipment in the klystron gallery. The perpetrator had dug a tunnel under the fence near the injector and had placed and detonated the device. Damage was estimated at $45,000; the accelerator was not operating at the time, and the planned turn-on in January was not delayed. No one was hurt. The FBI investigated to no avail.

The problem we faced was that the perpetrator did not send us a message, so the purpose of the attack has to remain purely speculative. At the time, due to budget pressures, there were many disappointed job applicants and a desperate one among these may have retaliated. Because we were making the auditorium available to anti-war protesters, someone from the right may have objected to such activities. Conversely, someone from the left may have carried out the attack because SLAC represents a large governmental installation. To this day we do not know the motive for this attack.

I was also drawn into the conflict by addressing large crowds in one of Stanford's outdoor arenas. When student protesters proclaimed they were going to "put their bodies on the line," I was able to calm them down by suggesting that being students, their minds were more valuable than their bodies. In that connection, I participated in organizing what was called "Student Workshops on Problems of Social Interest" (SWOPSI), and taught a SWOPSI course on arms control. I am pleased that at least one of the students in that course became a senior arms control officer in the State Department.

But the most serious crisis stemming from the student unrest paralleling the directorship of SLAC was generated by my election by the faculty to the advisory board of the university. Normally, that board simply reviews nominations for tenure appointments to the faculty and then makes recommendations to the provost and president. However, the advisory board is also charged with adjudicating under "due process" any controversy in the very rare cases where the university administration decides to terminate a professor's tenure. That conflict became real when President Richard Lyman proposed to dismiss H. Bruce Franklin, a tenured professor of English and a scholar expert in the works of Melville. Franklin was charged with disrupting a talk given in one of Stanford's auditoriums on January 11, 1971, by Henry Cabot Lodge, the conservative American ambassador to the United Nations. He was also charged with leading a student assault wielding axes which resulted in breaking into the computer center on February 10, 1971, which was believed to run computer programs in support of the Vietnam War. In addition, Franklin was charged with two other episodes of inciting students to riot and engage in vandalism. As a result of these charges, the advisory board held hearings from September 28, 1971, until mid-November of that year. The hearings went on six days a week from noon to 7 PM. There were seven of us, together with a secretary, and a legal counsel who was a member of the UC Berkeley law faculty. The board was known as "Snow White and the Seven Dwarfs;" the secretary was Snow White and I was Doc.

Due to the formal judicial nature of the proceedings, I was in a difficult position. While these proceedings were going on, the president was formally a "Party" before our board, and therefore, if I would talk to President Lyman in the absence of Professor Franklin or his counsel, this would be a forbidden "ex-parte" contact. On the other hand, I reported to the president in my role as director of SLAC. Happily, there were no crises at SLAC during the period of the hearings, so that this conflict did not lead to problems.

There were violent episodes associated with the board hearings themselves; a hearing was interrupted by a group performing a revolutionary skit on "People's War" and the home of one board member suffered an arson attack. As a result, the university provided all-night guards at the homes of all board members. The guard at my house precipitated a nightly conflict with our

Australian sheepdog; the guard was afraid of dogs and the dog took exception to nightly prowlers. That episode ended with the dog giving up its inbred herding and watching instinct for the time being by being tied up for the night.

The hearings raised fundamental issues. Franklin maintained that the rules of conduct for the faculty were not explicit in forbidding his conduct, and therefore, the university's interpretation in censuring that conduct was "overbroad" and hence unconstitutional. On the other hand, if there were detailed rules (such as explicitly forbidding faculty members from breaking into the computer center with an axe), then the rules of conduct for the faculty would fill many volumes.

But the university went to extremes in the opposite direction by claiming that as a private university, Stanford could interpret rules of conduct for the faculty in a more restrictive fashion than would be acceptable in a public university such as the University of California. The board, after extensive discussion, rejected both these extremes. However, the board had to hear in detail the evidence on the facts concerning Franklin's alleged conduct.

Here again, the university's case was not convincing in all cases. Two elderly ladies testified that they saw Franklin personally and vehemently incite the students to interrupt the speech by Cabot Lodge. I found out where the ladies and Franklin were sitting in the auditorium and examined the sight lines in the auditorium and came to the conclusion that the ladies could not possibly have seen what they testified to have seen, perhaps excusable in the excitement; so the board dismissed that one charge. The other charges involved "incitement to unlawful conduct" and we were exposed to learning about past legal rulings on that subject; the standard was that such conduct to be unlawful must be both threatening and imminent, which was the case here.

After lengthy hearings in which Franklin represented himself but was obviously previously advised by legal counsel, but where the university was directly represented by counsel, the board contemplated its verdict and wrote its decision. In the middle of the writing process, we received a telegram by an eminent Harvard law professor wishing to intervene as a Friend of the Court. Our learned legal counsel recommended that we should send him an obscene telegram with a note "rude letter to follow." We decided to ignore the intervention.

The board's decision, five to two, found Franklin culpable on three of the four counts as charged and recommended dismissal. This decision became the subject of a brief law course at Harvard University. It was later challenged in court and upheld with the exception of one finding of fact bearing on the illegal conduct. A later university advisory board, whose members had not been participants in the hearings, had to rule on that matter. Professor Franklin was dismissed and accepted a position at Rutgers University.

In summary, the activities not relating to SLAC or its research were quite diverse in the period following completion of construction of SLAC, but the team in place at the laboratory was perfectly adequate to proceed in the absence of the director.

13
Fixed Target Research at SLAC

Although the research at SLAC after its completion was guided by proposals for specific experiments, it is no accident that the initial program largely consisted of exploitations of the facilities just built. It is also noteworthy that the research period after completion of SLAC was managed along the organizational lines that were used during construction. We maintained the four principal divisions: administrative services, business services, research division, and technical division. The technical division made a seamless transition from managing a large construction project to operating the accelerator. That division, through a newly established experimental facilities department, provided engineering support to the multitude of experimenters.

Research groups were formed in the research division headed by individual members of the faculty. Initially, I personally participated in the work of Group A, headed by Richard Taylor. However, I only co-signed the first paper emanating from Group A on elastic electron scattering, essentially extending the previous work of Robert Hofstadter to much higher momentum transfers. The results[1] showed no surprises, but continued the momentum transfer dependence of the electric and magnetic form factors along the pattern established previously. After that, taking into account my administrative and outside commitments, I reluctantly decided that it would become impossible for me to remain an active participant in primary experimental work; thus all of my subsequent publications were either conference reports; summary or policy papers on high-energy physics topics; papers on design, construction, or general principles of large accelerators or colliders; or publications dealing with activities outside high-energy physics.

SLAC commenced research during what might be called the second high-productivity period of high-energy physics. Let me make some brief and very incomplete and sketchy remarks on the status of particle physics when SLAC research began.

In previous chapters, I commented on the first exciting period at LBL after the war, where the properties of pions discovered in cosmic rays in 1947 became manifest. The pion work continued, largely at Chicago and

Columbia, establishing in greater detail the character of pion–nucleon scattering and the structure of pion–nucleon resonances.

But then came the wave of discoveries of "strange" particles, first discovered in cosmic rays in 1947 as "hooks and forks" or "V-tracks," and then pursued further, principally at Brookhaven and LBL. At the time, "strangeness" was considered to be an additional quantum number conserved in strong, but not weak, interaction and was not initially associated with a new constituent of existing or conjectured particles. The spectroscopy of strange particles was pursued extensively and resulted in the systematic arrangement of many new strange particle states in what was dubbed "the eightfold way." In turn, that systemization gave rise to the conjecture by Gell-Mann and Zweig that these states were combinations of "quarks" of charge $\pm 1/3$ or $\pm 2/3$. This resulted in an interesting but ultimately fruitless debate as to whether these quarks were purely mathematical descriptions or whether they corresponded to physical reality.

Then came other parallel exciting developments. It was shown that the decay of what was first believed to be two distinct neutral K mesons appeared to result in final states of opposite parity, leading to the conjecture that two such mesons existed at essentially the same mass but of opposite parity. This puzzle was resolved by T. D. Lee and C. N. Yang, who proposed that these two particles were one and the same, but that parity was not conserved in the weak interaction. This conjecture was confirmed by experiments showing that ordinary radioactive beta decay from spin-aligned nuclei exhibited an asymmetry consistent with the parity nonconservation picture.

But then Fitch and Cronin showed that the long-lived K_2 particle could decay with a small branching ratio (about two parts in a thousand) into two pions, which was the dominant decay mode of the short-lived kaon. This experiment, which has since been performed with much higher precision, could only be explained if not only parity (P) but also the combination of parity and charge conjugation symmetry (C), that is, CP, is also violated in the weak interaction. The question remained at the time whether the violation of CP symmetry is restricted to the kaon system.

Following these revelations, the existence of the neutrino, which had been proposed by Pauli to explain the continuous beta-decay spectrum, was experimentally discovered in 1956. The μ meson, or muon, remained a principal outsider in any kind of consistent pattern either in cosmic rays or resulting from pion decay. The muon could no longer be considered to be the carrier of the nuclear force, as had been proposed by Yukawa. The pion had replaced that role and cosmic ray experiments had demonstrated that the muon was not strongly interacting. I. I. Rabi made the famous remark about the muon, "Who ordered that?" Adding to this experimental pattern was a conjecture by Iliopoulos in 1970 that in addition to the strange quark, another quark dubbed "charmed" might also exist and if so, would resolve then-extant problems in strong interaction theory. But the existence of a fourth quark as yet had no direct experimental confirmation.

SLAC entered into this picture when there was an enormous mass of accumulated data, interpretation, and theoretical work in elementary particle physics, but no single model unifying all this had appeared on the scene. The SLAC research program, which in the initial proposal had aimed at the much humbler goal of simply extending the early electron-scattering results at the MARK III accelerator at Stanford's High Energy Physics Laboratories and to serve as a source of secondary particles, was to play a major role in converting this rather unsatisfactory picture at the end of the 1960s into what today is called the Standard Model. The history of this exciting period to follow is well documented in *The Rise of the Standard Model.*[2]

I only give some selected accounts of these experiments here; they were the work of many individual physicists carrying out experiments at SLAC. Let me start with the electron-scattering work using the three spectrometers in End Station A shown in Figure 13.1.

The taking of data flowing from the diverse detectors was greatly facilitated by the deployment of computerized recording and analysis, largely designed by the MIT component of the collaboration. Although this experiment had processing capacity comparable to that in a 21st-century washing machine, at the time its computing capacity was a pioneering accomplishment.

Initially, the program using these instruments was designed to extend elastic scattering of electrons from the proton and the neutron (in the deuteron)

FIGURE 13.1. The spectrometers in End Station A rotating about a common pivot. (Credit: SLAC photo.)

to higher energies, and then to extend this work to inelastic scattering, leading to the known "resonances" or excited states of the nucleons. But the attention of the experimental groups was rapidly deflected to examining the "deep inelastic scattering" (DIS) which left the nucleons fragmented in a continuous set of energy states. The reason for this attention was simple: the DIS cross-section turned out to be much larger than had been previously surmised. In fact, the ratio of DIS, as a function of the momentum transfer to the nucleons, to that of scattering from a charged point particle, showed only a very slow variation (Figure 13.2), which is in contrast to the very steep decrease with momentum transfer exhibited by elastic scattering.

Deep-Inelastic Scattering and the Discovery of Quarks

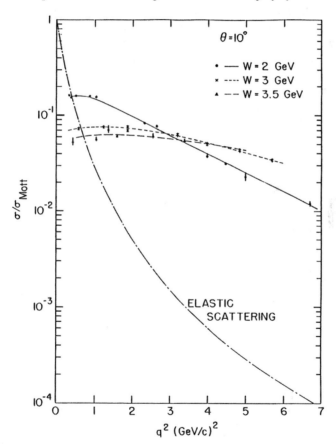

FIGURE 13.2. The ratio of deep inelastic scattering (DIS) cross-section of electron scattering in hydrogen to the theoretical Mott scattering cross-section from a point charge, plotted as a function of the square of the four-momentum transfer. The elastic scattering cross-section is plotted for comparison. (From Hoddeson et al.,[2] Fig 32.2.)

These spectacular results called for innovative interpretation. Many theorists and the experimenters themselves became engaged in trying to understand these results, however, the crucial insights came initially from J. D. Bjorken ("BJ") and then from Richard Feynman. BJ showed that the data exhibited "scaling," meaning that the "structure functions" of the electric and magnetic components of the DIS cross-section could be expressed as a function of a single variable x, which is the ratio of the square of the momentum transferred to the nucleons divided by the difference in energy between the incident and scattered electrons. In turn, this scaling behavior (Figure 13.3) corresponds to the kinematics expected if the scattering took place from individual pointlike particles within the nucleons, "Raspberry seeds within the jam," in the words of Sid Drell. These subparticles, dubbed "partons" by Feynman, were subsequently, and with increasing confidence, identified with the fractionally charged quarks conjectured by Gell-Mann and Zweig in 1964.

I had the privilege to present the preliminary data and their conjectured interpretation at the Fourteenth International Conference on High Energy Physics in Vienna in 1968. I said "Theoretical speculations are focused on the possibility that these data might give evidence on the behavior of pointlike charged structures in the nucleons."[3] Richard Taylor, Henry Kendall, and Jerome Friedman received the 1990 Nobel Prize in Physics for this work, which established the foundation of the physical reality of the quark components of the Standard Model.

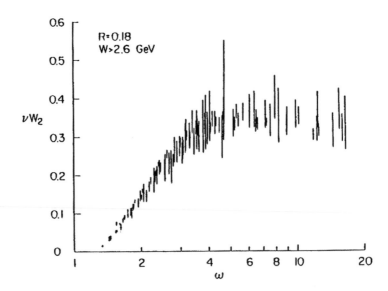

FIGURE 13.3. The "form factor" of DIS is plotted against the "scaling variable" ω which is the ratio of the energy loss of the scattered electron to the square of the four-momentum transfer. (From Hoddeson et al.,[2] Fig. 32.3.)

Needless to say, this account paints a highly superficial picture of the much deeper insights that resulted from these experiments. In turn, the measurements of deep inelastic lepton scattering have been continued through further work at SLAC, DESY, and at the Jefferson Laboratory, and through experiments using muon beams.

Before giving an account of the experiments at the other SLAC facilities constructed during the establishment of the linac, I describe further developments in End Station A. Burton Richter and his group conducted a series of photoproduction experiments there using the large spectrometers, which gave important but not surprising results. Moreover, an important development was the establishment of polarized electron beams to be used in scattering experiments in the End Station. The use of polarized beams at SLAC was initiated by Vernon Hughes of Yale University. He had developed a polarized electron source at Yale that consisted of a polarized atomic beam of Lithium 6 which was then photoionized with a flashlamp, separating the polarized electrons. He proposed this apparatus as an electron source for SLAC, and that proposal was accepted. However, I was very concerned about the possibility of accidentally introducing lithium into the accelerator. If this happened, it would probably seriously limit the gradient that could be sustained by the linac. (Vernon accused me of being paranoid about lithium contamination, and I agreed with his accusation.)

The experiments proceeded successfully with a number of automatic shut-off devices being provided between Vernon's electron source, dubbed "PEGGY," and the accelerator. A number of experiments were executed with the polarized beam on polarized targets. These gave interesting results, but not of sufficient precision to establish what is now known as the spin paradox, which shows that the spin of the proton and neutron are not predominately carried by the primary quarks in these nucleons.

The big advance in polarized beams was initiated in 1978 for a crucial experiment by a group led by Charles Prescott. He was interested in establishing the interference between electron scattering due to the exchange of a photon with that of exchanging the carrier of the weak interaction. Because the weak interaction had been established to be parity-violating, the expected result would be an asymmetry in the scattering cross-section as a function of scattering angle on an unpolarized target as a function of the spin orientation of the incident beam. At the time Prescott initiated his experiment, conflicting parity-violating results had been achieved by atomic spectroscopy experiments, and Prescott's experiment hoped to clarify this picture. He succeeded in greatly increasing the sensitivity for parity-violating effects in electron scattering by a number of measures. Primarily, his group produced a new electron gun that generated longitudinally polarized electrons by photoemission from gallium-arsenide photocathodes illuminated by circularly polarized laser light. In addition, the group used some of the magnets of the existing spectrometers to provide a wider aperture magnetic analyzer at a fixed angle.

The resulting experiment was highly successful and gave me a great deal of pleasure in that it was a successful combination of efforts using theory, the electron injector, the accelerator itself, the beam analyzing system between the accelerator and the End Station, and finally, the detecting spectrometer.

At the time of the experiment, Weinberg and Salam had developed their electro–weak theory, which integrated the theory of electromagnetic inter-actions with that of the weak interactions. One of the implications of that theory was the existence of "neutral currents" in the weak interaction, meaning that the weak interaction should not only be involved in beta decay, which involves the exchange of an intermediate charged object, but should also participate in reactions where a neutral object is exchanged. The existence of the neutral current had been demonstrated at CERN through neutrino interactions in a heavy-liquid bubble chamber; Prescott and his group hoped to confirm the existence of neutral currents by demon-strating quantitatively the parity-violating properties in electron scattering, and thereby measuring the critical parameter of the Weinberg–Salam theory, which defines the degree of "mixing" of the electromagnetic and weak interactions.

The development of the gallium-arsenide source was an important achievement in its own right, and that type of source has become the stan-dard in further polarized electron beam developments. It replaced the atomic beam-based sources and its degree of polarization increased from the roughly 50% available for Prescott's experiments to the 90% range available today. On a personal note, Willibald Jentschke, the director of DESY, spent a sabbatical at SLAC, joining Prescott's group and taking charge of one of the rapid polarizing devices in the laser beam illuminating the cathode, the so-called "Pockels cell." We enjoyed his company as a personal friend.

The Prescott experiment dramatically showed the changes in scattering intensity as a function of the polarization state of the incident electrons. That state could be flipped rapidly using the Pockels cell, but it could also be changed by changing the energy of the incident beam. Because that beam was being deflected in the beam switchyard, the polarization of the beam changed as a function of energy by the so-called g-2 factor, which defines the precession of an electron in a magnetic field. Figure 13.4 shows the scattering intensity as a function of energy, dramatically showing the change as a function of that precession. I recall the occasion when Prescott presented his results in a semi-nar at SLAC. The results were so cleancut and persuasive that when the chair of the seminar asked for comments and discussion, there was a deadly silence. No one could think of any possible questions on the validity of the results. This classical experiment confirmed the basic correctness of the Weinberg–Salam electro–weak theory, which is now an essential component of the Standard Model.

It was much easier to define the program on the facilities at SLAC built dur-ing construction than it was to make decisions among the many proposals of

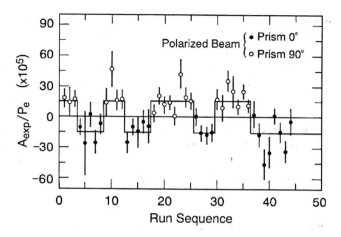

FIGURE 13.4. The left–right asymmetry in scattering of longitudinally polarized electrons from deuterium. The data are displayed as the polarization is reversed by rotation of the Pockels cell prism. (From Hoddeson et al.,[2] Fig. 27.3.)

the "dog that did not bark" type, that is, searches for new phenomena that might or might not have a firm basis in theoretical predictions. I now outline a number of experiments in this category that were carried out.

A comparison between electron and positron scattering was carried out in End Station A. Such an experiment would be sensitive to higher-order interactions, that is, scattering in which more than one photon is exchanged. No statistically significant differences between electron and positron scattering were observed; of course, the intensity of positron beams available at the time was considerably lower than that for electron beams. An experiment on a possible asymmetry between positive and negative muon production by photons was carried out by Mel Schwartz from the Stanford physics department, again with a negative result. Then a hunt for possible long-lived leptons heavier than the electron was undertaken by Martin Perl. Much later, he indeed did discover the tau lepton, the heaviest member of the lepton family, which turned out to have a much shorter lifetime than could possibly have been found in these early experiments. Perl also conducted experiments comparing electron and muon scattering, again to find signals elucidating the difference between electrons and the "mysterious" muon. No statistically meaningful differences were found.

Then there were a number of "beam dump" experiments where the electron beam was stopped in a heavy absorber and searches were undertaken hunting for particles emerging from the absorber. One possible source of such particles would be neutrinos interacting in the absorbers, which then would generate secondary particles that could be detected. Such an experiment was carried out by Mel Schwartz without surprising results, although he expressed strong unhappiness at the time that a higher-density absorber was not made available that possibly would have made neutral current events accessible in sufficient quantities to be significant.

In addition to these experiments, there was a very productive program in which I had no involvement that was carried out using the various secondary beams in End Station B and in the central beam. I have already mentioned the continuation at SLAC of the converted 82-inch bubble chamber from the earlier Berkeley program. This proved highly productive in many further studies of hadron spectroscopy. There were extensive photoproduction studies in the 40-inch rapid-cycling bubble chamber, and also in the large streamer chamber constructed by Robert Mozley. In addition, David Leith constructed a multilayer spark chamber to study high-energy pion and kaon interactions.

This personal account does not recite the results from these experiments here; this omission of course does not reflect on the experiments' importance, which was very high indeed.

14
New Facilities—Colliding Beams

Well before the construction of SLAC was completed, there was interest in constructing facilities in addition to those planned for in the initial complement of research equipment. Specifically, the interest in colliding beam facilities at SLAC had already started in 1961 in discussions involving Burt Richter, David Ritson from the physics department, and others. As noted previously, the electron–electron colliding beam facility at HEPL did not start generating results until SLAC construction was well under way. It was well understood at that time that the research program which an electron–electron collider could support was much more limited than what an electron–positron ring could do, but at the same time it was recognized that electron–positron rings required additional development and of course, a positron source. In the meantime, interest in electron–positron rings arose in Europe. A small electron–positron ring called ADA was completed in Italy in 1961. It functioned as designed but operated at too low an energy to support research. A larger ring, ADONE, was completed in 1967 at 1.5 GEV per beam and rings of similar energy were constructed in Orsay, France, and in Novosibirsk.

The discussions at Stanford culminated in a proposal that I strongly supported, submitted to the AEC in 1963 in preliminary form, and finalized in 1964. At the same time, a proposal for a large electron–positron collider was originated by the Cambridge Electron Accelerator, and in response, a committee to compare the two proposals was established by the AEC. It was led by Jackson Laslett, and it came out in favor of the Stanford initiative. Notwithstanding this recommendation, the proposal to construct an electron–positron collider at SLAC was not approved for construction in the years 1965, 1966, 1967, 1968, 1969, and 1970! Stemming from these successive frustrations, I reached an agreement with the controller of the AEC, John Abedessa, for SLAC to go ahead and build an electron–positron storage ring anyway—without separate construction authorization—by diverting equipment funds from the regular SLAC budget. In parallel with these developments, the Cambridge Electron Accelerator laboratory succeeded in constructing a "bypass" that deflected electrons and positrons from the electron synchrotron into a sidetrack where collisions could occur.

Notwithstanding this complex history, the SLAC electron–positron ring, at a maximum energy of 3 GEV per beam, turned out to be probably the most effective particle collider ever built as measured by its productivity in relation to its cost. The original design consisted of two asymmetric rings. The particles were to be stored in two separate pear-shaped tubes and were to undergo collisions at two interaction points. For this reason, the machine was called SPEAR, an acronym for Stanford Positron–Electron Asymmetric Rings. The design was then changed to a single symmetric ring, but the name SPEAR stuck. For this effort, Burt Richter gathered together a group of experienced accelerator physicists, including some from the Cambridge Electron Accelerator. This generated somewhat of an anomaly in that there was now a team of "circular" accelerator physicists in the SLAC research division, whereas the "linear" accelerator physicists remained in the technical division. Under the very restricted circumstances under which SPEAR was ultimately constructed, this arrangement turned out to be satisfactory.

SPEAR was constructed without a formal building; rather the enclosure consisted of concrete shielding blocks to provide shelter. There was not even a concrete floor; the machine was erected in the SLAC target area with the magnets supported on piles consisting of pipes penetrating the asphalt pavement of a former parking lot. Despite the relatively primitive civil engineering, the machine incorporated many sophisticated technical features: a radiofrequency system, an ultra-high vacuum system, and inflection channels in addition to the magnetic lattice itself. Actual construction started in October 1970 and the first beam was stored in April of 1972.

In parallel with building the machine, the team collected by Richter constructed a detector to be installed in one of the interaction regions of SPEAR. That detector, later called the MARK I to distinguish it from its successors, incorporated a solenoidal magnetic field, tracking chambers, and a total absorption calorimeter, providing a very large solid angle for detection. The detector was built in collaboration with physicists from the Lawrence Berkeley Laboratory. As it happened, the MARK I detector became the prototype for all future detectors for colliding beam machines. It was deliberately designed to be nonspecific as far as particular reaction channels were concerned. Rather, the patterns from the large majority of collisions were to be recorded as they occurred, with the isolation of specific reaction channels to be accomplished offline. This arrangement proved to be exceedingly successful, taking account of the fact that the reaction rates were limited because, of course, the density of the colliding beams is much lower than that of stationary targets. In order to maintain flexibility, I made the decision from the beginning to open the second interaction region to proposals designed to examine specific reaction channels. That decision turned out to be unwise. Experiments were indeed carried out in the second interaction region: one experiment emphasized secondary particle production of energies higher than were accessible for particle identification in the MARK I. Another experiment carried out by Robert Hofstadter and his group used sodium iodide detectors for gamma ray identification. Due to their finite solid angles, neither of these experiments attained counting rates sufficient to be competitive with the much more universal MARK I detector.

One of the primary objectives of SPEAR was to investigate the rate of production of hadrons resulting from electron–positron annihilation. The ratio, called R, of hadron production to electromagnetic pair production processes had been investigated previously at some of the other storage rings built abroad and in Cambridge. All those early results produced R values considerably larger than had been theoretically predicted, and therefore a detailed investigation of the R value at the higher energy accessible to SPEAR was strongly indicated. The earlier experiments had explored the annihilation of electrons and positrons into pion pairs, and had identified the strong peak of such production at the so-called ρ resonance. Beyond that, the R value remained large, but the results from the different machines were quite widely scattered.

A large part of 1973 was spent improving the software on the MARK I detector in order to improve the identification of electrons, muons, and hadrons. When this process was completed, an energy scan of the R value was started in 200 MEV steps. As late as July, 1974, Richter reported in a London conference on high-energy physics that the R variation as a function of energy was smooth and that the cross-section for hadron production was roughly constant in the energy range between 2.5–4.8 GEV. Good agreement with measurement at the other laboratories was reported. But in October, 1974, a reanalysis by Roy Schwitters exhibited some disturbing deviations from the smoothness, so, starting in November of 1974, it was decided to search the region near 3 GEV in 10-MEV steps in order to investigate a suspicious bump near 3.2 GEV which amounted to a factor of 2. Further scanning in that region had shown some variability, now known to have been caused by inaccurate energy setting.

Then on Sunday, November 10, 1974, the November Revolution started. A peak was seen in the scan using small energy steps which were two orders of magnitude above the value predicted by pure quantum electrodynamics! Clearly, the earlier bumps that had triggered this reinvestigation in small steps were due to the radiative tail of this extremely sharp peak, shown in Figure 14.1. I was called in the early morning of Sunday, November 10 and went to the SPEAR control room. All I can remember is that I was walking around stammering, "My goodness! My goodness!" being incapable of more coherent comments.

On the evening of November 10th, Sam Ting arrived from MIT for a previously scheduled meeting of the SLAC Program Advisory Committee. During that evening, Ting heard about these spectacular results, and on the morning of November 11th, a meeting was held in my office attended by Burt Richter, Sam Ting, and myself.[1] Ting showed the results obtained at MIT, which saw a similar peak but with inferior statistics in the spectrum of electron pairs resulting from hadron collisions. At that meeting, it was decided

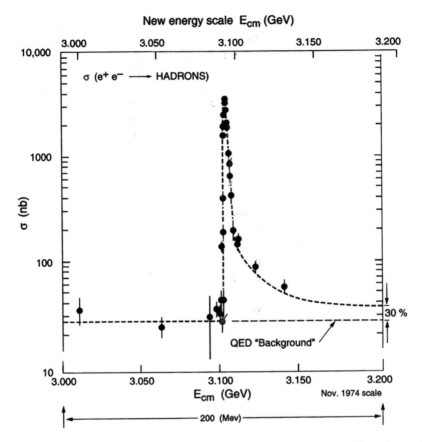

FIGURE 14.1. The sharp peak in hadron production as a function of electron–positron collision energy observed at SPEAR during the November Revolution of November 10, 1974. (From Hoddeson et al.[2])

that Ting and the SLAC group would announce these findings simultaneously and would also synchronize their publication. A seminar was held at SLAC in the afternoon of Monday, November 12th, where Schwitters announced the SPEAR results and Ting revealed the MIT data. Gerson Goldhaber gave a seminar at LBL at the same time. A manuscript was immediately prepared for publication by both groups and flown East on Tuesday, November 13th. The news of these spectacular events spread rapidly.

On November 15th, the ADONE machine in Frascati, Italy, confirmed the peak by pushing their machine a slight amount above its designed limits.

At a later date, I thanked John Abadessa, the AEC controller: "I would like to report the discovery of an unauthorized particle on an unauthorized colliding beam facility."

A note might be in order to discuss the meaning of "discovery." The MIT group had for several months been collecting data on the electron pair spectrum

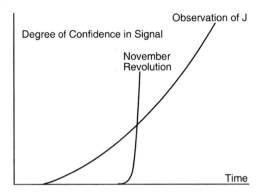

FIGURE 14.2. The ambiguity of the meaning of "discovery." The degree of confidence in a phenomenon is plotted as it evolves over time during a "gradual" ("Observation of J") and "surprise" (November Revolution) experimental observation. (From the Panofsky Collection.)

in this energy range from stationary targets bombarded by protons. As the data accumulated slowly, the statistical significance of their peak increased accordingly, and alternate explanations for the peak were being considered intensely. Therefore, the confidence which Ting and his collaborators had about the reality of their peak grew only slowly, and he restricted discussion of his data to a small group of colleagues. In contrast, the confidence in the peak found at SPEAR jumped from nonexistence to total confidence in a matter of hours on November 10. Figure 14.2 sketches how the confidence plotted versus time progressed in the two experiments. Thus, the priority of discovery depends on what level of confidence is considered adequate for public disclosure. In some respect, it was the secrecy accompanying the slow growth in confidence at MIT that led to this situation.

I was a witness to the developing events at SLAC and participated in the decision that led to the joint announcements and simultaneous publications of the MIT and SPEAR results. Because of the continuing discussion as to the relative priority of the two independent results, I published a note[3] in *Science* of which I quote the last paragraph here:

There is no question that the Massachusetts Institute of Technology–Brookhaven National Laboratory discovery represented a very difficult and superbly instrumented piece of work in high-energy experimental physics, and the authors deserve full credit for that achievement. Similarly, the independent LBL–SLAC discoveries represented a spectacular demonstration of the powers of electron–positron storage rings in discovering new particle states and in exploring the spectroscopy and intrinsic properties of such particles. This should be a joyous occasion for all physicists.

The spectacular discovery and the November Revolution caused some immediate publicity. I believe the first newspaper covering the result was the *Daily Californian*, the student newspaper at the University of California,

Berkeley. Ting and Richter persuaded the editors of *Physical Review Letters* to publish both the MIT and the SPEAR observations in the same issue of December 2, 1974, and thereby to make an exception to their usual policy of refusing to publish any results that had previously appeared in the news media.

The November Revolution was followed by a detailed experimental exploration, in 1-MEV steps, of the entire energy spectrum available to SPEAR, with each step usually taking only three minutes. At the same time, attempts were being made worldwide to understand the nature of these results. Although it took some time, the consensus emerged that the 3.1-GEV peak, now called ψ/J, was a combination of a charmed quark and its anti-particle. On November 21, a second peak called ψ^1 was discovered at 3.685 GEV and a whole spectroscopy of the charm–anticharm or "charmonium" system emerged. The discovery of the charm–anticharm states raised the obvious question whether combinations of the charmed quark with other quarks identified earlier could be observed as specific particle states. This question was answered in the affirmative by the detailed work of Gerson Goldhaber and his LBL collaborators, who showed that a peak of kaon and pion combinations of effective mass 1.87 GEV could not be explained in any manner other than as being a combination of a charmed quark and a quark of another flavor. Many investigations led to that definite conclusion, including the yield of this particle state as a function of colliding-beam energy.

The combined discovery of the charmonium states and what was now called "the naked charm" states—incorporating only a single charmed quark—led to a worldwide cottage industry in developing the spectroscopy of this new family of states. This included work not only at SPEAR, but also at DESY in Hamburg and at other European laboratories.

Burt Richter and Sam Ting were jointly awarded the Nobel Prize for their discoveries that had been announced in consequence of the November Revolution. Those discoveries and the follow-on activities were succeeded by a second Nobel Prize-winning discovery, which added another building block to the Standard Model. However, the dynamics of this second discovery were different from those of either Richter or Ting. The discovery by Richter's group exploded almost overnight, whereas Ting's painstaking work resulted in progressively increasing evidence, which was shrouded in secrecy until the revelation of that evidence was triggered by the November Revolution. The history of the discovery of what is now called the tau lepton was quite different.

Martin Perl, the leader of that effort, became interested in the nature of the leptons immediately after joining SLAC in 1963. In his earlier work, Perl questioned why the electron and muon were identical as far as their interactions were concerned, but different in their mass. As noted previously, his initial experiments at SLAC, which attempted to distinguish between electron and muon interactions, gave negative results, as did his search for not-as-yet-known leptons of long-enough lifetime to be discovered in external beams. Perl's group and Richter's group were joined in the early experiments on SPEAR together with the group from the Berkeley laboratory, and searching for heavy leptons was part of the original proposal for the MARK I detector. Searches for such leptons had been carried out at the Italian electron–positron storage ring ADONE, but had succeeded only in establishing limits.

After the November Revolution, Perl noted in early 1975 that there was an excess of electron–muon coincidences beyond what could be expected from accidental overlaps or misidentification of particles. Perl revealed the existence of these anomalous electron–muon events in a talk in June 1975, but was careful to note that these events could be of multiple origin; for instance, they might be decays from other known particles, either mesons or nucleons. But of course, the possibility that there was an additional lepton heavier than both the μ and the electron was a leading candidate for the most plausible explanation.

Such a particle was expected to decay either into muons or electrons accompanied by two neutrinos, or alternately, could also decay into one neutrino and a member of the hadron family. The heavy lepton interpretation was demonstrated to be more likely by the addition of the so-called muon tower to the MARK I detector, which could trace muons more precisely to higher energy. But there was difficulty in identifying the decay channel into pions plus a neutrino. As late as 1977, at the Lepton–Photon Conference in Hamburg, the status of such events was still in some doubt, although the threshold behavior of electron–muon coincidences was reported by the DESY experimenters. But then the cloud lifted.

The decay of the heavy lepton into electrons was identified at another detector (DELCO), and in 1978, Perl was able to announce that the evidence was now fully convincing that the existence of a third member of the lepton family, which he called the tau lepton, was now fully established, and that the decay modes of such a particle were behaving close to what was theoretically predicted for a "sequential" lepton. Of course, by now the tau meson is firmly embedded into the Standard Model: more precise measurements of its mass, lifetime, and branching ratios have been carried out in many laboratories.

These major discoveries, as well as the systematic measurements that followed, amply justified the effort which went into the construction of SPEAR. The MARK I detector was replaced with a device (the MARK II detector) using similar operating principles but improved components, and Robert Mozley added an additional detector, the MARK III, to further add precision to the exploration of the tau–charm region.

At about this time, I was faced with the decision whether to dedicate some of the running time of SPEAR to using the x-rays copiously produced from the electrons circulating in the storage ring as tools for a wide area of applications. The initiative to do so came from Stanford's applied physics community, and we initially decided to dedicate the x-rays emitted by SPEAR to experiments in various fields of applied science in a "parasitic mode," that is, using such beams while the machine was being run for high-energy particle physics. For this purpose, it was agreed to establish what was initially called SSRP, the Stanford Synchrotron Radiation Project. SSRP was not part of SLAC, but was operated as an independent laboratory under the aegis of the associate provost for research of Stanford University. I had been quite reluctant to incorporate SSRP into the SLAC structure because I felt that SLAC being a single-function laboratory was a source of strength in that a single branch of the government, the high-energy physics office of the division of research of the AEC, would have to assume full responsibility for the health of SLAC.

SSRP was highly successful in its own right, in fact, so successful that the arguments for its growth and eventual incorporation into the SLAC structure by my successor (Burt Richter) became persuasive; I do not describe here this evolution. It involved construction of many beam lines, the construction of several insertion devices to generate x-rays of higher brightness, and an arrangement to share beamtime on an equal basis between synchrotron radiation physics and high-energy physics. Eventually, it was decided to dedicate SPEAR entirely to synchrotron radiation photon research. The resulting laboratory, now called SSRL, has been and continues to be a highly successful part of SLAC's operation. Initially, when SSRP started growing, I made the observation that we had to distinguish between "symbiotic" and "parasitic" operation of the synchrotron radiation program at SPEAR with high-energy physics activities. Symbiotic means that the functions can exist together, whereas parasitic means that the parasite kills the host. In some respects, that latter pattern prevailed, but with highly productive results.

I was often asked, after the initial completion of SLAC construction, how long the laboratory could productively operate. My standard answer was: "Ten years, unless someone produces a good idea." Now, four decades later, the original accelerator is still running; there were lots of "good ideas," in addition to SPEAR.

Many other initiatives for construction activities at SLAC were proposed by members of the community, but were not as successful as SPEAR had been, or did not come to fruition at all. One initiative was the so-called Recirculating Linear Accelerator (RLA). The idea was to take the beam from the linac, store it in an external racetrack in parallel with the machine, and then re-inject it into the linac, resulting in a beam of twice the original linac energy. I supported this effort, and a design report was submitted in August 1973 after a planning effort that had started in 1971. The proposal was turned down by the Atomic Energy Commission as being premature, considering the productivity level that SLAC was then achieving without this addition. In retrospect this was probably not a bad decision, and we did not appeal it. As it turns out, we were able to increase the energy of SLAC by other means later, but I note that the idea of recirculation of a linac has become a valuable addition to the arsenal of accelerators and has been exploited in particular by the Jefferson Lab in Newport News.

Another major project that was initiated aimed at converting SLAC's linac to a superconducting accelerator. A feasibility study to do so was carried out under the leadership of Perry Wilson, and he projected that eventually an energy near 100 GEV could be reached through such a conversion. This projection was based on the theoretical limit set by the magnetic field maximum that would quench superconductivity and which was expected to occur at a gradient of about 30 million volts per meter. The problem was that the voltage gradient attained in practice in superconducting accelerating structures at that time was less by an order of magnitude.

Experiments on the Stanford campus at the High Energy Physics Laboratory aimed to convert the old MARK III accelerator into a superconducting machine, but those efforts failed to attain a gradient even close to expectation, and the program was eventually terminated without leading to a practical machine. After the design study at SLAC, I carried out a review and decided to cancel this project, to the great disappointment of those members of the SLAC staff who had dedicated a large effort to the design study. In retrospect, indeed, that decision was the correct one at the time, inasmuch as it took almost three decades (!) to develop the superconducting RF technology to a level that made it competitive with room temperature linear accelerators. In fact, to jump ahead in time, superconducting RF technology has now become the technology of choice for the International Linear Collider, which is the most important and most ambitious next step visualized today in accelerator construction.

An immediately successful approach to increase the energy of the SLAC accelerator was the SLED (SLAC Energy Doubler) project. In the original proposal to establish SLAC, an expansion option was identified to quadruple the total klystron power by increasing the total number of klystrons and their power sources accordingly. That option was never exercised. Rather, we decided to increase the peak microwave power into the accelerator by pulse compression, that is, compressing the pulse length from the modulators into a much shorter pulse and increasing the peak power correspondingly. This conversion was highly successful and made it possible for the present SLAC linac to eventually reach an energy of over 50 GEV, of course, at the cost of sacrifice in duty cycle.

15
International High Energy Physics

Particle physics has always been a subject of international interest, and that interest is widening. Because of its remoteness from direct application, international communication on the subject has been relatively unimpeded throughout periods of political tension, including the Cold War. Throughout my responsibility for SLAC, I always strongly encouraged international participation in SLAC work; the standard language at SLAC, in particular on the third floor of the Central Laboratory where the theoretical physics group resides, is broken English.

Our cooperative interactions with Europe were generally informal rather than being implemented through negotiated agreements. SLAC staff members who were on leave frequently spent time at CERN, the European international laboratory. However, while I was director of SLAC, formal bilateral agreements were negotiated on a governmental level with the U.S.S.R., China, and Japan, and bilateral committees were established with each of these countries. In addition, the International Union of Pure and Applied Physics (IUPAP) established a subcommittee on high-energy physics. At various times I served on each of the bilateral committees, as well as on the IUPAP Subpanel. This service involved innumerable trips and meetings that I do not enumerate here. Rather, this chapter describes some of the specific initiatives in which I was involved, that frequently went beyond the cooperative activities sponsored by these committees.

Following the groundbreaking, or rather "Iron Curtain-breaking" visit by American physicists to the U.S.S.R. in 1956 described in Chapter 11, cooperation with the Soviet Union continued. The Soviets participated in the recurrent conferences on both high-energy physics and on instrumentation for high-energy physics which were organized under IUPAP auspices. Actual Soviet experimental contributions were relatively moderate, and my own interest focused particularly on working with the Institute of Nuclear Physics at Novosibirsk in Siberia. I developed a real friendship with Gersch Budker and Alexander ("Sasha") Skrinsky, and there were frequent visits. The Budker Institute of Nuclear Physics (BINP), as the Institute in Novosibirsk was renamed after Budker's death in 1977, continued its pioneering work in colliding beam technology and also contributed a great deal to plasma physics. However, exploitation of the various pioneering colliding beam devices for particle physics remained relatively moderate.

In 1973, Budker and I initiated discussions for a more intense collaborative effort aimed particularly at mutually reinforcing their pioneering instrumental work with the greater experience in particle physics experimentation here at home in the United States. In June of 1973, after careful preparation of supporting memoranda, Burt Richter and I had discussions in the U.S.S.R. with Budker, his chief experimentalist V. A. Siderov, and Skrinsky to start a formal joint project. We considered four alternatives: a joint project in a third country; a joint project at Stanford; a joint project at Novosibirsk; or a joint project divided between Stanford and Novosibirsk. After considering the political situation, we decided to concentrate on a potentially major collaborative project at Stanford, which we specifically decided would be joint construction and operation of a 15-GEV per beam electron–positron storage ring. We also decided to collaborate later on further research. We had lots of discussion on the proposed division of effort, to be designed in such a way that there would be no actual transfer of funds between the United States and the U.S.S.R., but there would be sharing of "people effort" and of instruments constructed at the home laboratories.

Richter and I, together with Budker, made a formal presentation of this proposal to Mstislav Keldysh, the president of the Soviet Academy of Sciences, and Moisey Markov, the head of the physics branch of the academy, in a meeting in Keldysh's office. The presentation to Keldysh on this joint enterprise was received cordially, but both Keldysh and Markov were noncommittal during the meeting. However, subsequently, these joint activities were vetoed by the Soviets, and Budker was instructed by Markov to formally recant the joint proposal. It was a sad occasion, but perhaps in retrospect, not too surprising. Such collaboration was probably premature; however, had it succeeded, it would have set an important precedent.

SLAC and the Novosibirsk laboratory continued friendly relations after this fiasco, but material contributions were restricted to business transactions: hardware was constructed by BINP and then acquired by SLAC through normal procurement procedures. Later, the Novosibirsk laboratory made a heavy commitment to participate in the work of the ill-fated superconducting supercollider laboratory in Texas (SSC); I describe this effort in a later chapter. Collaboration with the Russian laboratories and the work of what is now called the Joint Consultative Committee continues today, but nothing as ambitious as the initiative proposed with Gersch Budker has been subsequently attempted.

My involvement with China has been much more productive. I was visited in 1973 by Zhang Wenyu, the leading physicist in China interested in high-energy physics. He had previously spent some time in the United States as a member of the faculty of the University of Minnesota, where he had done cosmic ray physics, and where he had supervised the Ph.D. thesis of William

A. Wallenmeyer, who later became the director of high-energy physics (HEP) for the U.S. Atomic Energy Commission. Zhang then returned to China and became a senior member of their Institute of Atomic Energy. In 1973, Zhang and an accompanying group visited U.S. laboratories preparatory to counseling their government on how China might establish a major facility in high-energy physics. China subsequently decided to build a 50-GEV proton synchrotron near the Ming Tombs outside Beijing. I was always critical of that proposal, because building such a machine and its accompanying laboratory would be very expensive, and its energy would be below that of both the National Accelerator Laboratory (now Fermilab) in the United States and the machines at CERN. Considering these facts and the magnitude of the effort required for successful experiments, it would be unlikely that China could make real contributions.

I had expressed some of these views to Zhang, and subsequently he invited me to visit China for a period of discussion on the subject. My visit took place in 1976, and I went together with my wife, Adele. It was a very eventful, two-week trip. The Chinese hosts were exceedingly gracious and invited us to an extensive post-Beijing tour that included stops in Guilin, Guanzhou, Shanghai, and Nanjing. I gave numerous talks and was treated as a VIP guest, but was generally somewhat isolated from most local contacts. Adele was shown many additional sights and paleontological exhibits and institutes.

As it happened, our visit followed the disastrous earthquake, centered near Tangshan, which also had done a great deal of damage in Beijing. Many buildings were damaged and declared too dangerous for occupation, and therefore the People's Liberation Army (PLA) had provided tents for people to live in along many of the main thoroughfares. I had inquired whether this disaster should lead to postponement of my trip, but Zhang firmly maintained that the importance of high-energy physics was such that we should proceed as scheduled.

So we came and had many discussions with Zhang and other physicists. During that visit, I suggested that an electron–positron collider would be a much better initial venture for China, because such a machine could serve a dual purpose of "serving the economy," by being a facility for synchrotron radiation, while at the same time allowing them to enter a field that was just beginning to be explored in the West. In contrast, I suggested that entering elementary particle physics via the Ming Tomb machine was like trying to jump on a train moving at high speed.

Aside from the scientific interactions at several institutes, the Chinese Academy invited us to visit several parts of China. This was dually interesting. First, we had never visited China before, but at the same time, we happened to be present at a time of major political upheaval. Chairman Mao Zedong had died, and so had Zhou Enlai, and a power struggle ensued between the so-called Gang of Four, which included Mao's widow and the new leadership headed by Prime Minister Hua Guofeng. Wherever we went, there were posters plastered on almost any available vertical surface. Some of these attacked the Gang of Four with lurid pictures of their projected fatal disposition; but there were also posters relating to more local conflicts. During periods set aside for "a short rest" in our official itinerary, we frequently escaped from our assigned quarters and took an extensive series of photographs of these posters; this collection turned out to be of great interest at home. Our guides were always ambiguous as to what all this was about, until the actual arrest of the Gang of Four. Then, after we departed from Nanjing on October 22, 1976, the guides were allowed to explain it all to us on a "now it can be told" basis. This was a very eventful trip and we provided a great deal of documentation on it.

Following this China trip, it was arranged for Fang Yi, the vice premier of the Chinese State Council, to visit U.S. laboratories, including SLAC and Fermilab. During these visits, there were extensive discussions on the options for China to pursue in HEP, and Bob Wilson, the director of what is now Fermilab, supported the idea that China's best start would be an electron–positron collider. After Fang Yi's return to China in 1979, Deng Xiaoping and President Jimmy Carter signed the official United States–China Agreement on Cooperation in Science and Technology at a ceremony that I attended. This agreement covered cooperation in science and technology in general, but the first protocol annexed to that agreement was in high-energy physics, and was signed by Fang Yi and by Jim Schlesinger, who was chairman of the AEC at that time. Under this agreement, a Joint Committee on Cooperation in High Energy Physics has been meeting every year since then. Following these consultations, the Chinese government agreed to sponsor the construction of the Beijing Electron–Positron Collider (BEPC) to be the centerpiece of the Institute of High Energy Physics.

Construction of BEPC involved extensive collaboration with SLAC in many forms. The Chinese sent a delegation of about 30 engineers and physicists to SLAC in 1982 to make a preliminary design of the machine. We housed this group in an upper story of what was then the Central Control Building, and from there they could summon various members of the SLAC engineering and physics staff for consultation about their design.

This visit had its amusing moments. We found rooms for the group in surrounding communities, where small numbers of the visitors lived together with several members of the delegation occupying one apartment. They would arrive at SLAC every day on bicycles and wearing their Mao suits, and they generally led an austere existence. As it happened, during that time the union contract of the Stanford University blue-collar workers expired, and a strike occurred. I vividly recall the image of 30 Chinese in Mao suits, on bicycles, crossing a picket line of American strikers, with both parties wondering what was going on. I gave a lecture to the Chinese emphasizing that this was a form of capitalist bargaining, with our workers wishing to sell their effort to the university at a higher wage. They took eager notes. The strike was soon settled, and life continued as if nothing had happened. Indeed, a complete preliminary design of the BEPC was produced during that summer's effort.

Subsequently, the Chinese authorized construction of the BEPC and extensive cooperation with SLAC continued. A group of several Chinese engineers were in continuous presence at SLAC, with the lead person having authority to spend foreign currency for procurements. At the same time, the Chinese made me an offer to be a consultant to their Academy of Sciences with substantial compensation. I refused that proposal, not wishing to be a consultant to a foreign country. Instead, arrangements were made to have my services be included in the annual agreements between the U.S. Department of Energy (DOE) and the Chinese Academy of Sciences, with the DOE "furnishing" me as a consultant without compensation. In this way, the arrangement could be vetoed at any time by me, by the Department of Energy, or by the Chinese Academy.

Under this arrangement, I made frequent subsequent visits to China, usually in order to remedy bottlenecks that arose during construction of the BEPC. The project manager for construction was Xie Jialin, who earlier had received his Ph.D. from Stanford University at the Microwave Laboratory. Xie had become a microwave expert while at Stanford. After receiving his Ph.D., the U.S. government did not permit him to go home for several years. He went instead to Chicago and oversaw the construction there of an electron linear accelerator to be used for cancer treatment at the Michael Reese Hospital. This turned out to be a very successful installation. The United States then relented and let him go home to rejoin his family in China.

Construction of the BEPC was an extremely visible project in China, and it was supervised from a very high level. Deng Xiaoping personally wielded the shovel at the groundbreaking ceremony, at which I was present. The Chinese government established a "leading group," consisting of high officials, which not only had authority to oversee the project but which, as a practical matter, served as expediters in case difficulties were encountered. I recall that issues having to do with land use were handled by the vice-mayor of Beijing, and difficulties encountered with klystron production were expedited by the future minister of aeronautics. The chair of the leading group was Madame Gu Yu, who was a senior member of the Politburo, and whose husband was Qiu Chimu, the chief advisor to Deng Xiaoping on political fundamentals.

On one of my visits in 1989, the group asked me to a meeting in which they presented their version of events at the Tiananmen Square massacre, maintaining that the original student protesters had later been replaced by "troublemakers" from outside Beijing. I made it rather clear that I was not convinced. At the same time, the high-level participation of the leading group proved essential in removing all kinds of bottlenecks during the construction period. It was always amusing to me that the type of jobs that at SLAC would have been carried out by the purchasing department were carried out in China by what amounted to a cabinet officer.

Under this management arrangement, unusual by Western standards, construction of the BEPC proceeded and was highly successful. The BEPC indeed moved Chinese high-energy physics to a contributing, if not leading, role in experimental high-energy physics. A noteworthy contribution turned out to be the most precise measurement of the tau lepton mass, and the most accurate measurement at the time of the ratio of hadronic to leptonic production resulting from electron–positron annihilation. The laboratory also made important contributions to various measurements in the tau–charm region.

Madame Gu Yu visited the United States in 1982 and consulted with the Chinese delegation at SLAC. We arranged for her to take a trip to the Central Valley of California, because in China she had some responsibilities for agricultural policies. I remember taking her to view some enormous spread of orchards in bloom. Her surprised reaction was, "Where are the people?" Indeed, not a single human being was in sight, in contrast to the situation in China. We took Madame Gu Yu to a lunch at the Harris Ranch in Coalinga, where she also observed the "beef factory" where various groups of cattle were being fed a variety of diets to prepare them for a diversity of customers. We invited a professor from the University of California at Davis to join us for lunch. He explained to Gu Yu that in California, at the end of the 19th century, well over 90% of the inhabitants were engaged in agriculture, whereas the number now was between 3 to 4%. This recital was prescient of the situation now in China. Due to increased mechanization, much of the agricultural population of China has now become superfluous, resulting in a large migration of Chinese from the countryside into the cities, thus becoming an enormous source of inexpensive labor for the industrialization of the country.

My wife and I made a second visit to China in 1979, and were invited on a memorable tour that included many visits to historical sights in the arid west. We were accompanied during the entire trip by Yan Wuguang, one of the senior physicists at the Institute for High-Energy Physics. He is the son of Yan Jici, who was a well-known Chinese physicist who made most of his contributions in France in physical optics, and who returned to China as a much "venerated" scientist. The younger Yan had the title of professor, which was awarded directly by the Chinese Academy of Sciences rather than by the Institute or one of the universities. The system of awarding faculty titles and the authority for supervising Ph.D. students appeared to me a very esoteric arrangement that made it difficult to reward substantial scientific achievement. In general, the social structure of the research teams at IHEP was highly personalized, and made communication somewhat difficult. It is going to take some time to modernize that structure.

The success of BEPC I and its detector, the BES or Beijing Electron Solenoid, led to plans for further expansion of the high-energy physics program in China. The local BEPC group generated plans for a tau–charm factory, an electron–positron collider at about the same energy as that of the BEPC, but generating a luminosity two orders of magnitude larger. The proposed device would consist of two storage rings, one in the existing BEPC tunnel, and the other in a second, adjacent tunnel to be constructed so that the two would have a common interaction region. A feasibility study for that project was completed by the end of 1995 as a cooperative effort of several Chinese institutions, and I strongly supported the resulting proposal.

However, at the time, there was increasing debate in China about the relative value of pure versus applied research, and as a consequence the Chinese Academy of Sciences decided not to support this "pure research" project, but instead to put more resources into a third-generation synchrotron radiation facility in Shanghai, to be used mainly by industry for applied research. As a consequence, only minor improvements to the BEPC were approved. At the same time, it was clear from the experience at the existing BEPC that scientific support for a new synchrotron radiation facility in China was indeed strong, but that industry at the time had very little interest in it. The users of the beam lines at the BEPC were entirely from scientific institutions, not from industry. Once the decision had been made to build the synchrotron, there ensued a series of extensive debates about both the siting of the Shanghai facility and about its sources of support, and this conflict caused major delays.

The proposals for minor upgrades of the BEPC appeared to me to be very cost-ineffective, but then the accelerator group at the Institute of High-Energy Physics "reinvented" the tau–charm factory, as a face-saving measure, by designing a compact two-ring machine that could be fitted into the tunnel housing of the existing BEPC. The Chinese approved funding for the new machine, dubbed "BEPC II"—which promised essentially the same performance as the earlier proposed tau–charm factory—but only after a great deal of valuable time had been lost.

That machine, the BEPC II, is now under construction. It is hoped that it will generate its first collisions in the year 2007. In the meantime, the social structure of Chinese basic science has changed drastically. There is no longer a "leading group," but the laboratory pursues new construction under the leadership of the laboratory director, with the government carrying out reviews at various stages, not unlike the American pattern. There is a new generation of young and very able Chinese accelerator physicists, such that international participation is less essential but still highly desirable, in particular in respect to the experimental program. SLAC is no longer as heavily involved, excepting in the various advisory committees. However, it is hoped that the new machine will offer an opportunity for accelerator-based high-energy physics at a time when the total number of such machines is shrinking drastically in the West.

Throughout the approval process in China for BEPC II, the joint United States–China Committee on Cooperation in High Energy Physics continued to meet annually, and we strongly supported the proposal for this initiative. On October 14, 1998, Premier Zhu Rongji was briefed during the 29th meeting of the U.S.–P.R.C. Committee. I gave a prepared talk on the role of BEPC II in high-energy physics and Premier Zhu retaliated by giving the visitors a lecture on the geology of China.

Interestingly enough, this briefing, which was reported publicly in the *China Daily*, was quoted in a January 3, 1999 report from a U.S. congressional committee chaired by Congressman Christopher Cox under the heading of "PRC Theft of United States Thermonuclear Warhead Design Information," an absolutely absurd reference. The Cox report was guilty of many other blatant misstatements and inaccuracies. I participated, together with some experts on United States–China relations,[1] in a review of the Cox Committee Report. Our critical review received significant attention and the Cox committee staff claimed that we disagreed only on interpretation, not on facts, a statement which again was contrary to fact. My interaction with China also continued in the area of international security and arms control: I refer to this in a subsequent chapter.

In 1991, a formal agreement between Japan and the United States on collaboration in high-energy physics was signed at SLAC. Japan had established a strong program in high-energy physics based principally at its KEK laboratory at Tsukuba. That laboratory had constructed a proton synchrotron followed by a "photon-factory" for synchrotron radiation and an electron–positron storage ring. And KEK has now become a friendly competitor with SLAC in that both laboratories are pursuing the exploitation of colliding beam facilities of very similar characteristics to study B-meson physics.

I have made numerous visits to Japan and over the years have developed a close friendship with Tetsuji Nishikawa, the founding director of KEK, and also with Satoshi Ozaki, who later returned to Brookhaven to direct its Relativistic Heavy Ion Collider.

During one of the visits to KEK we were hosted by Shinichiro Nakayama, the grandson of the secretary of the Japanese Cabinet who had been a close colleague of my grandfather who had worked on modernizing Japanese law, as described in Chapter 1. Shinichiro and I have maintained continuing grandson-to-grandson contact ever since.

Akihito, the Japanese Emperor, and his wife, Empress Michiko, paid a visit to SLAC on June 23, 1994. By that time I had retired as SLAC Director and Burt Richter was the official host, but I participated mostly to help in greeting the emperor and to talk physics with the Japanese graduate students who were invited to join the party. Plans for the imperial visit were carefully made but went seriously astray when a SLAC staff member accidentally locked the key controlling the elevator for the interaction hall of the SLAC Linear Collider (SLC) inside the elevator cab. It was planned that Richter would greet His Imperial Majesty on the ground floor of the hall and present some gifts to him but this proved impossible, leading to a reshuffling of the ceremonies. Michiko Minty, a SLAC machine physicist, was among those waiting for the imperial couple. When the empress reached Minty, she stopped and carefully studied Minty's name badge. The Empress asked about Minty's first name, and our Michiko responded, "My mother named me for you." Many jokes were made about a high-tech laboratory being unable to make an elevator run, but the United States–Japanese collaboration proceeded undisturbed.

International collaboration was active with many countries other than those that signed formal agreements for collaboration in high-energy physics. In parallel with the successful operation of the MARK III accelerator at HEPL, the French established a laboratory near Orsay in 1955 called the Laboratoire de l'Accélérateur Linéaire (LAL). Operation for research started in 1962 at about 1 GeV electron energy. The Stanford Laboratories were not directly involved in collaborative design but many of the prime movers in construction of that facility and in later research there had been visiting physicists at HEPL and SLAC. Conversely, SLAC physicists, including Richard Taylor, spent time at LAL during its formative years. I also worked closely with Maurice Levy, a prominent French theorist, who became science advisor to the French government, and who served for some time at the French embassy in Washington D.C.

A notable event in our relations with the French was the visit of President Georges Pompidou to SLAC in 1970, accompanied by Levy and others. Greg Loew, who was then the head of accelerator physics and instrumentation control at SLAC, and who is bilingual in French and English, accompanied Pompidou throughout and explained the layout of SLAC while the two were riding in a helicopter which later landed on the SLAC lawn. Security in connection with this visiting head of state was extensive. There were protesters on SLAC grounds expressing sympathy for Israel and objecting to the previous sale of Mirage fighter planes by France to Libya. Pompidou, who had an academic background, was extremely interested in details of the SLAC installation; and during his visit he also insisted on having an independent meeting, not attended by university officials, with a group of Stanford student protesters. Except for that occasion, I was a direct participant in all phases of the Pompidou visit. I presented a paper during the formal discussion between the French and U.S. groups on the difficult financial support situation for high-energy physics at that time, both in the United States and Europe. Overall, it was a lively and enjoyable visit by a head of state.

Cooperation was also very intense throughout the establishment and operation of the Deutsches Elektronen-Synchrotron (DESY) in Hamburg. Willibald ("Willie") Jentschke, the founding director of DESY, was a professor at the University of Illinois before returning to Germany and was very well acquainted with American high-energy physics. He visited SLAC many times, and he took a much-deserved sabbatical leave at SLAC after DESY was completed. Again, as was the case in connection with LAL, several of the prime movers in the creation of DESY had spent time and carried out research at SLAC prior to assuming their roles at DESY.

The foregoing account highlights some of the events in which SLAC played a key role in furthering international collaborative work in high-energy physics during the time while I was director. Although such events were indeed noteworthy and exciting occurrences, international collaboration is a fact in the daily routine of the laboratory and remains so to this day.

16
Advances in Accelerator-Based High-Energy Physics

SPEAR was an enormous success and, as indicated earlier, possibly the most cost-effective, productive high-energy physics machine ever built. It demonstrated the power of electron–positron colliders: several additional machines were built abroad and in the United States. SLAC's next step was PEP, an electron–positron collider employing two beams of a maximum of 18 GeV each. Construction of PEP was a collaboration between SLAC and the Lawrence Berkeley Laboratory (LBL). Ground was broken in June, 1977, and its operation started in 1980. Other than supporting its approval and construction wholeheartedly, I had very little to do with the project as such. It was carried out under the direction of Burt Richter and his "circular" accelerator physicists, led by John Rees. Because of the success of SPEAR, demand for experimental opportunities at PEP from other institutions was large and SLAC approved the construction of five independent detectors in separate interaction regions on it. A similar machine, called PETRA, was built at DESY in Germany.

One of the contributions of PEP was that it served as a testbed for improvements in detector technologies. The Time Projection Chamber (TPC), invented by Dave Nygren at LBL, received its first large-scale application at PEP. In addition, a very large superconducting solenoid weighing 107 tons was transported from Chicago to SLAC to provide the magnetic field for the so-called High-Resolution Spectrometer (HRS). The transport of this enormous coil was a major challenge because of its nearly 15-foot diameter. That challenge was amplified by the fact that the trucker who carried it was a publicity hound who advertised the transport of the coil to most of the communities the truck passed. These announcements were not only in words, but were also conveyed by the fact that the radio communication preceding the large truck was carried out on frequencies that kept activating automatic garage door openers in many homes along the magnet's path. The coil arrived in a blast of publicity at SLAC on November 23, 1979. All five PEP detectors reached operation as designed, after some initial difficulties.

As is often the case, PEP was to some extent an anti-climax to SPEAR. Although several very important measurements were made there, PEP generated no discoveries even remotely matching in excitement those of the November Revolution of 1974. Possibly its most important measurement established that the B quark lifetime was much longer than had been predicted, and this in turn made it possible to observe events in which the primary vertex-generating B-mesons could be separated from the secondary vertex characteristic of B-meson decay. A large number of additional data were generated, contributing to better understanding of tau meson decays, charmonium and charmed meson spectroscopy, and the beginning of B-meson spectroscopy. The three-jet events that provided evidence of the gluon, the carrier of the force among quarks, were discovered at PETRA in this energy region. But in retrospect, PEP was over-instrumented, as measured by the number of detectors installed in the multiple interaction regions relative to the poverty of major discoveries of the energy region it made possible for exploration.

The productivity of PEP was considerable, however, it was not as large as had been hoped for, because the energy region covered by PEP was bereft of qualitatively new phenomena. It was clear that an electron–positron collision energy near 100 GeV would be extremely desirable to explore the existence of the "intermediate boson" that was predicted by the Weinberg–Salam theory, which unified the electric and weak interactions. However, an electron–positron storage ring-based collider obeys a scaling law which indicates that the cost of an optimized design increases with the square of the collision energy, the reason being that the synchrotron radiation loss per turn varies as the fourth power of the energy divided by the bending radius. If you assume that the cost of building a very high energy electron–positron collider is derived from the sum of the cost of items varying linearly with the radius of the machine, and costs proportional to the average radiofrequency power required to compensate for synchrotron radiation loss, then both the radius and the radiofrequency power should increase as a square of the collision energy. Thus a 100-GeV electron-positron collider would be both expensive and space-consuming.

Such a machine (LEP) was constructed at CERN, but at SLAC Burt Richter proposed the construction of what we first called a "single-pass collider," meaning that electron and positron beams of about 50 GeV each should be directed at each other, collide once, and then be discarded. Ideally, such a device would imply having two linear accelerators of 50 GeV aim at each other, but Richter proposed an intermediate solution in which the paths of both electrons and positrons produced by the existing two-mile accelerator would be bent into "one-time" collisions. The required magnetic bends would, of course, generate synchrotron radiation, but because such bends occurred only once per linac pulse, rather than continuously, as is the case in a storage ring, the energy loss from synchrotron radiation is acceptable.

Unfortunately, synchrotron radiation energy loss is not the only detriment of such bends: quantum fluctuations of the synchrotron light emission also dilute the density of the beam both in space and energy, and this dilution grows rapidly with energy. However, calculations indicated that these problems are tractable at beam energies of 50 Gev, but that they would clearly get out of hand if attainment of substantially higher energies were attempted. Thus, higher energies would require a "real" linear collider using two separate linear accelerators, rather than the double-bend arrangement, shown in Figure 16.1, that was built at SLAC.

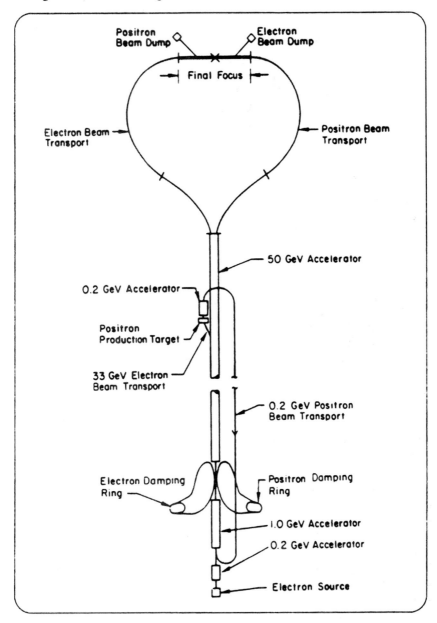

FIGURE 16.1. Layout of the SLAC Linear Collider (SLC). (Illustration by SLAC InfoMedia Solutions.)

FIGURE 16.2. Size comparison between the SLC and the LEP storage rings. (Illustration by Greg Stewart/SLAC Info Media Solutions. Photo credit: CERN.)

The SLAC Linear Collider (SLC) began construction in 1983 and provided its first collisions in 1989. The housing for the SLC required the excavation of tunnels separate from PEP, and only a single interaction point was practical. However, the resultant installation was much smaller and less expensive than the competing LEP machine at CERN; Figure 16.2 shows the relative sizes of the two installations.

Neither the excavations for PEP nor for the SLC tunnel led to any significant paleontological discoveries. However, my wife was "volunteered" to educate the construction workers to be sensitive to any fossil remnants that might be uncovered. For that purpose, she prepared a tray of fossil bones to show to the excavation crews. When carrying that tray past the guards at the radiation gate one day, she explained her purpose and was answered with the remark, "I didn't know there was anything dead out there!"

Construction of the SLC was a major undertaking. The numerous magnets in the two arcs were manufactured in-house at SLAC. They had to be of high precision considering their small aperture. The accelerator using the pulse compression scheme described earlier had to operate steadily at 50 GeV both for electrons and positrons. In addition, SLAC designed upgraded-power klystrons capable of delivering about 60 MW of peak power. We solicited bids from industry to build this newly designed tube but none were satisfactory to us. We therefore rejected all bids and built and tested the full complement of tubes (245 in all) in our own shops.

Above all, to obtain high-enough luminosity, the size of the beams at the interaction point had to be in the micrometer range in both dimensions. This, in turn, required that the positrons and electrons should be damped in separate "damping rings" at low energy (1.2 GeV) by reducing their radial excursions through synchrotron radiation. Moreover, the beams emerging from the two magnetic arcs had to be focused to the required small dimension by a magnetic lens system that had to be designed for very low aberration. In view of all these difficulties, it took a long time to commission the SLC to reach performance competitive in luminosity to that of colliding-beam storage rings, but success was eventually obtained.

The first detector at the SLC was the time-honored MARK II detector, whose performance had been demonstrated at SPEAR and PEP. However, to do justice to the new opportunities, a new detector, the SLD—incorporating many new ideas—was designed under the leadership of Marty Breidenbach. The interaction hall was designed to make it possible to alternately insert either detector into the single interaction region.

Completion of the SLD detector was interrupted by the great Loma Prieta earthquake in 1989. Happily, we had provided for the possibility of such an event by installing earthquake damping cylinders to restrain the motion of the major elements of the detector so no damage was done, other than squirting oil from the dampers across the floor and scaring the installation technician who was hanging from a crane at the time.

The earthquake did relatively little damage at SLAC, thanks to its initial conservative design which substantially exceeded then-applicable earthquake codes. Earthquake damage was also limited by an earthquake safety committee which for years had been making a major nuisance of itself by regularly circulating around SLAC reminding people to tie down tall bookcases, remove high loads from electronic racks, and so forth. The main effect of the 1989 earthquake was that "SLAC lost its bearings," meaning that the location of all survey monuments had shifted. The alignment of the accelerator stayed pretty well intact for the first 20 sectors. However, an offset of nearly one centimeter occurred about 100 meters from the end of the accelerator, and in addition, the SLC magnets required realignment. Thus the net damage in financial terms of the Loma Prieta earthquake was fairly small; possibly as much as two million dollars. This was in conformance with the original design principle which stipulated that SLAC be built so that any earthquake damage would be limited to less than a million in 1962 dollars.

SLC did great physics. It indeed observed and measured the large peak in electron–positron annihilation cross-section at the mass of the neutral intermediate boson that had been predicted by the Weinberg theory. Using the MARK II detector, after it had been moved over to the SLC from SPEAR, the width of that peak was determined with sufficient accuracy to provide the initial evidence that only three families of neutrinos could exist at energies below 100 GeV. In addition, the so-called Weinberg mixing angle was determined with the highest accuracy of any measurement by a single detector; the total accuracy generated by combining all the measurements taken by the numerous detectors at the LEP collider at CERN with the SLC determinations was, of course, significantly higher.

A principal reason why the SLC, despite its lower luminosity, remained highly competitive with the CERN LEP facility was its use of polarized electron beams. The gallium-arsenide cathodes, whose development I mentioned earlier, were greatly improved by straining the gallium-arsenide lattice, thus removing the degeneracy between the two relevant states and raising the polarization of the photoelectrons from about 50% to close to 90%. These electrons could be accelerated and deflected into the final interaction region with minimum loss of polarization. Of course, polarized positron beams remained a gleam in the eye of physicists for future linear colliders. It was electron polarization that permitted the measurement of the relevant weak interaction parameters with accuracies comparable to those obtained at CERN with much higher luminosities.

In summary, quite apart from the excellent physics results obtained at the SLC, that machine served as a first installation demonstrating the potential of linear colliders; it also demonstrated the value of using polarized beams in such machines. All this gave a major impetus to the worldwide consensus that has now emerged that a large electron–positron linear collider is the most promising tool to sustain the future of accelerator-based high-energy particle physics, complementing the work made possible by the LHC and providing further discovery potential.

I retired as director at SLAC in 1984 but remained a member of the staff until 1989, when I took my second retirement. Thus I was still director during the establishment of PEP, but only acted in that role during the very beginnings of the SLC. In parallel with all these events I became involved with the ill-fated Superconducting Super Collider (SSC). The parameters for that machine were generated as a result of a number of deliberations by committees of the National Academy of Sciences and of the Department of Energy's High-Energy Physics Advisory Panel (HEPAP). At that time, a proton collider called ISABELLE was under construction at the Brookhaven National Laboratory, but was encountering difficulties in the construction of the superconducting magnets.

Because the parameters for that ring were not very exciting, HEPAP recommended that ISABELLE be terminated and instead a Central Design Group (CDG) be established, starting in 1984, to be charged with designing a machine to achieve collisions between two beams of 20 TeV protons. That energy level was chosen by theoretical considerations. It became clear at that time that the Standard Model had to contain a particle or particles which would generate an explanation for the spectrum of masses that were otherwise empirical numbers within the Standard Model. The masses of such particles were at the time poorly defined by theory or experiment, but the energy chosen was very conservative, being based essentially on consideration of the Unitarity Limit which demanded that such new phenomena must be uncovered at a collision energy well below 40 TeV. The SSC Central Design Group was established at Berkeley under the leadership of Maury Tigner, a very experienced accelerator and collider designer and builder at Cornell University. The CDG issued a conceptual design report for such a supercollider in March of 1986.

In the meantime, from 1977–1980 I served on the board of trustees of the University Research Association (URA), which was overseeing what is now the Fermi National Laboratory. At the time, that laboratory was beginning to operate the Tevatron with its superconducting magnets, albeit with considerable initial difficulty. In response to the initiatives resulting in the establishment of the Central Design Group for the SSC, URA appointed separate boards of overseers for Fermilab and the SSC, and I served as the chair of the SSC board from 1984 to 1993. URA submitted unsolicited proposals to contract for the construction of the SSC in 1987 and 1988. Although President Reagan approved construction of such a machine in January 1987, accompanied by the somewhat enigmatic "throw deep" comment, the DOE did not accept the URA proposals. Rather, the DOE was bombarded by both congressional and industrial voices suggesting that a purely academic organization would not be qualified to build or run such an expensive undertaking. As a result, the Department of Energy issued a Request For Proposals (RFP) to all interested parties, whether industrial or academic, for the construction of the SSC.

URA succumbed to these pressures, and substituted a "teaming" arrangement in its proposal, in place of serving as the sole contractor for building the SSC. "Teaming" meant that URA would be associated with two industrial contractors "experienced" in building big government-sponsored facilities, largely in the defense and reactor arena. The chosen teaming partners were Sverdrup Corporation for conventional construction and the EGG Corporation, which had previously managed major DOE activities, principally in the nuclear weapons area. To devise this teaming proposal, URA had engaged the services of Douglas Pewitt, who had previously served on numerous assignments in and out of government. As it turned out, the URA teaming proposal was the only one DOE received, and it was accepted in January of 1989. Responding to President Reagan's endorsement, Congress approved the project and provided the necessary funds in 1989 and, in subsequent actions, in 1990, 1991, and 1992.

After President Reagan's endorsement and congressional actions, about half of the states within the United States submitted proposals for sites that would be "just the thing" for the location of the SSC. Each of these proposals greatly extolled the virtues of high-energy particle physics and the SSC. A site selection committee was established by the National Academy of Sciences and reduced the list of candidate sites by a large factor. The final selection of the site at Waxahachie, Texas was made "in camera" by the DOE. After the site selection, the SSC became known as "a Texas project" and miraculously, the other proponent states dramatically lowered their interest in high-energy particle physics.

As part of its proposal, URA was required to name a director for the SSC. I participated in identifying candidates, discussing their qualifications, and corresponding extensively on the subject, however, the final selection was made by the URA President. Roy Schwitters, who had played a large role at SLAC during the November Revolution of 1974, had then become a professor at Harvard, and later the spokesman for the large CDF detector at Fermilab, was selected.

Construction of the SSC was, from the beginning, beset by serious conflicts and troubles that eventually led to its demise. The complete and definitive story of the rise and fall of the SSC has yet to be written, although many articles on the topic have appeared. The most complete of these have been written by Michael Riordan,[1,2] but the fundamental question of "Who killed the SSC?" will never be fully answered, because so many factors contributed. Let me recount only some of the difficulties that directly touched upon the work of the board of overseers.

One factor was growth in cost, both real and artificial, generated by design choices, inflation, and delays, some due to deliberate funding stretch-outs, and some due to factors internal to the SSC laboratory. This cost growth is well documented in the final SSC Report.[3] The Central Design Group had estimated a bare construction cost near $3 billion in its 1986 report. That sum did not allow for inflation and contingencies, and also did not include experimental equipment and operating costs for research and development before and during construction. These omitted factors were provided for in subsequent congressional actions. However, a substantial jump in cost occurred when the "orbitry" of the protons in the machine was recalculated in detail in 1990, principally by Helen Edwards, who was in charge of the accelerator group of the SSC, and by David Ritson from Stanford University.

Their detailed calculation demanded that the injection energy be doubled from 1 TeV to 2 TeV, and that the aperture of the bending magnets be increased by 20%. The resulting cost increase caused a major uproar within the DOE, Congress, and the board of trustees and overseers of URA. To me the choice was clear: the large increase in cost implied by these changes was technically fully justified but would endanger the project, whereas the chosen collision energy of 40 TeV was not sacrosanct. The matter was considered at a joint meeting of the board of overseers and the board of trustees of URA. At that meeting, I was outvoted 19 to 1 as the boards jointly decided to preserve the collision energy of 40 TeV. This decision was supported by several theorists in subsequent congressional testimony.

Indeed, reducing the energy at this point in time would have been difficult and would lead to delays, because to capture the savings by that reduction, the radius of the machine would have to be shrunk correspondingly, which, among other changes, would mean starting the land acquisition process over again, from scratch. Whether preserving the budget and reducing the energy would have saved the SSC at that time is one of those "what if" questions that will never be answered. However, today, after the demise of the SSC, the

Large Hadron Collider (LHC) at CERN—scheduled for completion in 2007 to attain a collision energy of 14 TeV (roughly one-third of that of the SSC)—provides the only definitely surviving "next step" in energy for high-energy physics. The LHC offers the only tool for the next generation of accelerator-based high-energy physics, at least until or unless the International Linear Collider, now under design, begins operation, which will happen sometime after 2017, if at all.

Aside from this cost growth, whose causes included the change of parameters discussed and other more mundane reasons, there were many other problems, some of them very fundamental. The joint teaming arrangement adopted by URA under external pressures was less than successful. Sverdrup Corporation assigned a group of unqualified engineers to the project and their continued lack of productivity induced the SSC management, under the directorship of Roy Schwitters, to terminate the arrangement with Sverdrup and replace that company with a more qualified joint venture.

The partnership with EGG was successful in some respects and unsuccessful in others. It was successful in that EGG was able to rapidly produce entire administrative departments—business services, accounting, procurement, personnel—"on a silver platter." This meant that Schwitters could direct more of his energies to the scientific aspects of the project, rather than to selecting individual leadership for administrative departments, which would have taken longer. However, the individuals made available by EGG were largely unacquainted with the research community and its idiosyncrasies. As a result, the administrators exhibited a great deal of insensitivity to research needs and were generally more interested in establishing and maintaining bureaucratic "law and order" in conducting their administrative functions, rather than in supporting the needs of the SSC research community. As a result, both the procurement activities and personnel operations were poorly matched to the needs of a laboratory whose ultimate output was to be basic research. It was just this situation that was successfully avoided in the early days of SLAC by having the administrative groups of the laboratory be designated as "services" rather than divisions, reflecting the fact that their functions were to support the laboratory's science.

A further problem, and possibly the most important one was that the then-Secretary of Energy, Admiral James Watkins, simply did not trust scientists to manage large technical enterprises; the extraordinary effectiveness demonstrated by scientists in the large and complicated World War II research efforts had apparently receded from memory. Instead, Watkins trusted his Area Manager for the DOE (Joseph Cipriano) as an "Experienced Administrator." Cipriano indeed had experience in managing construction of naval bases, but none in scientific laboratory construction or management.

At the same time, the ability of the SSC director to make independent decisions, in particular in the allocation of funds to the diverse line items of construction, was severely limited by the contract. There were literally thousands of such line items where DOE approval was required before funds from one line item could be moved to another. This mistrust by Admiral Watkins escalated to the extent that relatively minor management decisions at the SSC were communicated by Secretary Watkins to Cipriano, who then passed them on to Schwitters, resulting in a lack of the kind of flexibility that had served SLAC so well during its construction phase.

The impact of all this was severe. Relations between Schwitters and Cipriano deteriorated to what amounted to nonspeaking terms, and the influx of managers with experience in industry but no contact with the world of science led to significant societal schisms within the laboratory. Watkins insisted that Schwitters appoint a project manager to be interposed between Schwitters and the various division heads, theoretically reporting to Schwitters, but in practice receiving direction from the DOE. That project manager, an engineer named Ed Siskin, added to the length of the communication chain between the director and the scientific staff and Siskin frequently jumped the communication chain entirely by reporting directly to the board of overseers.

The biggest production item for the SSC was the acquisition of the superconducting magnets. The individual in charge of that acquisition was an experienced engineer in the Navy's Trident program, but again a person who was insensitive to science. He believed that the scientists should transmit their requirements "over the transom" and he would deliver to meet their needs. In other words, he essentially cut off any communication between the scientific community and his own engineering staff on how the technical work was to be accomplished, not a happy arrangement.

The scientific community maintained strong interest in the SSC, although the tensions enumerated here produced some prominent defections from participation. Two large collaborations evolved for constructing detectors, and combined financial support for these collaborations was provided by the state of Texas and the federal government.

A further critical obstacle was the issue of international financial support for the SSC. Although there was indeed substantial international participation, this aid came in the form of collaborating physicists, reimbursed services, or provision of accelerator or collider components that were purchased from abroad. In particular, the Budker Institute of Nuclear Physics dedicated a large fraction of its entire program to support the SSC. I remember that during my frequent visits to Waxahachie, I drove around with Sasha Skrinsky, who turned out to be more knowledgeable on the geography of Dallas, Texas than I was and who kept me from getting lost.

It was very difficult to bridge the gap between President Reagan's initial support for the SSC, justified as a means of demonstrating the technical "superiority" of the United States, and the congressional demand for international financial cost-sharing in SSC's construction. Europe in general—and CERN in particular—proved unresponsive to U.S. requests for financial contributions. CERN, being itself international, albeit regional, in nature, had problems meeting the financial needs for its own program and future aspirations.

Japan appeared more promising, and the matter of Japanese contribution was on the agenda of a summit meeting between President Clinton and the Japanese Prime Minister in 1993. I had a meeting with Vice President Gore, together with the Nobel Prize winning physicist Steve Weinberg, to brief the vice president prior to the summit meeting on the case for the SSC. I gave a talk on the technical features of the SSC, and Steve lectured on the present status of our theoretical understanding of the nature of matter, including the origin of our universe in terms of the Big Bang. Gore's reply was, "What happened before the Big Bang?" And Steve Weinberg pretended to know! A rather interesting meeting, but notwithstanding our presentation, discussion of the Japanese financial contribution "dropped off the end of the agenda" during the subsequent summit meeting, and such a participation never materialized.

This failure did not help in Congress, which exhibited increasing animosity to the SSC. As a more liberal Congress was elected, the conservative positions of the Texas congressional delegation became counterproductive; the Texans tended to vote against any social initiatives while at the same time emphasizing that billions of dollars were needed for the SSC. The House voted to cancel the SSC for the fiscal years 1992 and 1993, only to have the Senate vote for the project and then prevail in conference. However, that magic failed for fiscal year 1994, when the House refused to accept the Senate position, even though the House–Senate Conference went along with the Senate. The support by the newly elected Clinton administration was less than enthusiastic.

Congressional criticism went to some absurd extremes. The SSC director was accused of maintaining potted plants in his office at taxpayer expense. Criticism also targeted the fact that visiting committees, as well as the board of overseers, were "entertained" at restaurant dinners. Indeed, this was true, but in contrast to industrial practice, all visiting SSC committees and boards served without compensation. So the taxpayer was getting a bargain. However, the DOE secretary did not reject such accusations when they were raised, but gave only the general response, "It will never happen again."

So the project was cancelled in October 1993, and a short-term contract for fiscal year 1994 was negotiated to provide for orderly termination. The tunnel segments were sealed to prevent access that might endanger people, and the various material assets of the SSC Laboratory were either distributed to other DOE laboratories (including SLAC) or were sold.

In my view, cancellation of the SSC was not only a tragedy for American high-energy physics but for science in general. Although indeed the cost growth was significant, it was tiny relative to that experienced by the International Space Station, which was in competition with the SSC when the SSC was cancelled. The Space Station became an important political signal that the United States and Russia could work together in a cooperative project, but it has never done any science and has experienced serious technical difficulties. At the same time, the Large Hadron Collider at CERN also encountered serious cost increases and consequent delays in the acquisition of its superconducting magnets, which in turn put the squeeze on the balance of CERN's scientific program. Hopefully, the LHC will commence research in 2008 at one-third of the energy projected to have been reached by the SSC eight years earlier. The LHC provides for two major collaborations in support of its two large detectors; each collaboration has roughly 2000 Ph.D. members, including a substantial number of Americans. Although this large a number, or at least almost as large a number, can be justified scientifically, it signals a pattern that is not overly attractive to young students contemplating a career in physics.

Service on the board of overseers for the SSC constituted my last direct involvement in the efforts to establish new accelerators. Subsequent to my retirement, SLAC initiated and completed the conversion of the PEP collider into the PEP II or B Factory machine; that installation was an asymmetric collider in that it brought 9-GEV electrons and 3-GEV positrons into collision. As a result, the center of mass of the collisions is not stationary, and this motion produces measurable distances between the primary collision and the decay vertex. At the same time, a machine of almost identical characteristics was constructed in Japan and both colliders each established an international collaboration to construct and operate a detector (BaBar at SLAC, and BELLE in Japan). The productivity of these machines has been spectacular. Among other things, they established the violation of charge conservation times parity conservation (CP violation), or in more popular terms, the matter–antimatter differences in B-meson decays, and they measured the relevant parameters.

SLAC is taking a leading role in promoting the International Linear Collider, the next step in colliders, following beginnings of operation of the LHC, and is maintaining a major engineering design enterprise for that purpose. Rather than commenting here on these initiatives in detail, some general remarks on the status of accelerator-based high-energy physics may be in order.

I have been a participant in a period of absolutely drastic changes in the methodology of experimental particle physics. During the time of my previous work at Berkeley described earlier in this book, accelerators were completed largely by those physicists who were given the opportunity to design and execute experiments for their use. The experiments described earlier, which established the parities of the pi-meson and determined the mass difference between the charged and neutral pions, were literally designed in the stockroom. I searched around for any available vacuum tubes that might be used as the basis for amplifiers suitable for the multiple detector channels for the detector used. Experiments were designed to target one particular channel of production of new particles and one particular mode of their decay. All other information produced during the experimental run was thrown away.

This general approach has now all but disappeared. Increasing specialization has made it impossible for an individual to remain competent in accelerator technology and the facets of detector technology required to conduct an experiment, let alone to remain fully acquainted with ongoing theoretical developments that initially motivated the experiment. At the time of my early work, some experimentalists were also extremely competent theorists. Enrico Fermi, and, to a lesser extent, Jack Steinberger, are examples. Although I dabbled in theoretical endeavors in accelerator orbit dynamics and some of the theoretical underpinnings of the experiments themselves, I am certainly not a theorist. I remember Ed McMillan's famous remark expressed at a Rochester Conference: "Any experimentalist, unless proven a damn fool, should be given one-half year to interpret his own experiment."

Today's dynamics of the creation and practice of accelerator-based particle physics, in particular collider physics, are vastly different. The reaction cross-sections are generally much lower, roughly decreasing with the square of the energy. Colliders have very much lower luminosities, that is, reaction rate per unit of cross-section, than do accelerator beams striking stationary targets. As a result, the present-day experimentalist must be "greedy;" that is, he must record as large a fraction of what occurs in the collision as is possible. This is accomplished by surrounding the collision point with a detector that subtends as large a solid angle around the collision point as feasible and registers most properties of the collision event. Most of what one might describe as "physics" is then done offline through computer programs designed to isolate the processes of interest which with increasing collision energy are becoming a smaller and smaller fraction of the total number of events that actually occur.

This evolution means that the detector is really a composite of many detectors to register the charge, momentum (by bending in a magnetic field), velocity, total energy, and other characteristics in separate elements of the total detector. Thus the design and construction of the detector requires the work of many collaborating physicists, electronics experts, and mechanical designers. Moreover, the time to construct such a detector becomes comparable to that of constructing the collider itself, and the construction funds for the detectors are a significant fraction of those required for building the collider. SLAC was the first installation that undertook design and construction of detectors on a schedule paralleling that of the accelerator-collider itself.

This evolution of collider and detector construction methodology had to be paralleled by a corresponding evolution of the social structure of the organization accomplishing the construction, the experiment, and the theoretical analysis. Accelerators are being built by specialists both in the technical components of the accelerator or collider as well as by experts in orbit dynamics, vacuum and RF technology, and so forth. The design and construction of the detector is the job of a collaboration of an ever-increasing number of members dedicated to building the various "layers" of the detector, each of which is dedicated to a specialized function.

Software designed to isolate the events of interest out of an enormous mass of less-interesting data is becoming an ever-larger component of successful experimentation. Indeed, many participants in designing such software have never seen the detectors or the colliders that produced the events to start with (one of my grandsons is an example in that category). Thus the "critical" size of an experimental collaboration continues to grow, a phenomenon not restricted to elementary particle physics.[4]

In parallel with these developments, which are driven by the evolution of the science itself, there are other social developments that may or may not be as unavoidable. Although the cost per unit of collision energy generated has greatly decreased, the total cost of each new installation has increased as the energy frontier advances. As a result, the total number of frontier installations worldwide is shrinking, and the question of how large a collider can be "afforded" continues to be raised. "Affordability" is of course not an absolute concept, but is driven strongly by the political perception as to what fraction of the total of national expenditures can be dedicated to this area of most fundamental research. The present perception is that the next elementary particle collider, which has now become the International Linear Collider or ILC, cannot be "afforded" by any one country, or even one continent.

This appears indeed to be true as a matter of political reality, but it is certainly not true if one observes that the construction cost of the ILC, spread over nearly a decade, is comparable to the cost of a week of the present-day war in Iraq. Indeed, the ILC is now a worldwide enterprise and is being creatively pursued by an international effort dedicated to conceptual design and engineering, as well as to exploring the methodology for funding and site selection. Such international pursuit is hardly a consequence of "affordability," however, it is certainly a creative exercise in demonstrating the feasibility of a truly international undertaking, and it may set a valuable precedent for other endeavors in that respect.

The above brief outline of the evolving dynamics of accelerator-based elementary particle physics at the frontiers of energy indicates that this particular pursuit may be in its terminal phase. But to compensate for this fact, there is now a convergence of the exploration of the very large and the very small. Because the processes involved in the early epochs after the "Big Bang" involved energies that are addressed by today's colliders in the laboratory, results of laboratory elementary particle physics shed light on the early processes in the cosmos; at the same time cosmological observations provide evidence on elementary particle physics, or at least severely restrict the theoretical excursions that are possible in interpreting elementary physics data. Thus particle astrophysics and laboratory elementary particle physics are becoming converging disciplines, mutually reinforcing each other.

This convergence is in most respects a happy one, in that a transition in methodology is occurring while the basic thrust toward understanding the basic nature of inanimate matter remains the same. There are, however, some unfortunate consequences of this evolution in methodology that are worth mentioning here, and which could be ameliorated.

The relationship between science, particularly physics, and government has evolved from the partnerships during and after World War II to today's less balanced government–contractor relationships. The necessary increases in cost, combined with official amnesia about the achievements of physicists during World War II, have resulted in increasing pressures by governments sponsoring the work for expanded control, and for shifting from requiring "accountability" to requiring "compliance" with government orders and regulations by the physicists and other individuals responsible for executing the program. Such regulations, imposed in the interest of health and safety or designed to improve efficiency, frequently are counterproductive, because they are generally designed for missions other than doing science.

In deciding whether to approve a new elementary particle physics project, governments insist on the identification of more and more "deliverables," that is, definable, predictable products of the projected work. As a result, the task of my successor laboratory directors has become much more burdensome and complex today than it was during my tenure.

Government sponsors generally seem to have forgotten that in the past most accelerators and colliders (including those at SLAC) have had their maximum creative impact in areas quite different from those emphasized in the proposal for their construction! The time scale for the political process between initial proposal and eventual authorization has also greatly lengthened.

Another result of the current evolution has been a justified demand for prioritization of the proposals—or even ideas—under consideration. Such priorities are generally established by advisory committees, which are now not only composed of practitioners of the field in question but also include representatives of other disciplines, and even nonscientific personnel. Again, this appears well justified, inasmuch as society as a whole is paying for the enterprise, and is expected to benefit in the long run. However, such priorities, once established, tend to be interpreted excessively rigidly, with the result that the laboratory managers have little flexibility in allocating funds to new, small, but creative initiatives within their own laboratories. Moreover, existing smaller collider facilities tend to be terminated before they have exhausted their potential in producing still very valuable data, once such facilities reach a point in their existence where they are unlikely to produce further spectacular discoveries.

In consequence, the decentralized initiative, which has been the basis of most of the exciting developments and productivity related in this account, has been stifled to a significant extent under the current pattern. Let me reemphasize that some of this pattern is unavoidable, and is determined by the evolution of the science itself. But many of the administrative practices are leading to unnecessary restraints on the creativity of the field and are adversely affecting its productivity.

Notwithstanding the drastic shifts in methodology during my active role in elementary particle physics, that period has been an exceedingly productive one, in particular in solidifying the Standard Model. Let us hope that this productivity can continue in the new evolving environment!

17
Science and Politics After Retirement

In 1978 Senator Claude Pepper, who was then 78 years old, introduced legislation that made it illegal to retire federal employees based on chronological age, and this legislation was extended later to apply more generally. In consequence, I made a private agreement with my fellow associate directors of SLAC that we would retire at age 65 as bureaucrats and retire as members of the faculty, if applicable, at age 70. This agreement is of course not legally enforceable, but has remained a guideline thereafter, although not exactly obeyed. As a result, Joe Ballam and Dick Neal retired as associate directors in 1982, and I announced my retirement as lab director in 1984 and became emeritus professor in 1989. My letter to the president of the university announcing my retirement as director was delivered about a year in advance of the actual date; it indicated that SLAC was in good shape and that there was time for a formal search committee to be convened to designate a successor, but it was also clear that Burt Richter was an obvious candidate. Indeed, the president of Stanford appointed a search committee that selected Burt as my successor.

Retirement ended my formal responsibilities for SLAC affairs, but succeeding directors have generally encouraged retired directors and faculty members to continue their activities at SLAC. Both Joe Ballam and I availed ourselves of that opportunity although Richard Neal withdrew from his SLAC activities and resumed life anew in Southern California. The end of SLAC responsibility led to the concentration of my work fairly heavily on nuclear weapons arms control, in addition to various advisory and committee activities.

A focus for addressing security matters was the Committee on International Security and Arms Control (CISAC) of the NAS which I had joined in 1981, and chaired from 1985 to 1993. That committee was the successor to an earlier group headed by Paul Doty of Harvard University which had engaged in bilateral scientist-to-scientist contacts with a counterpart Soviet group headed by Academician Millionchikov. After Millionchikov's death, those bilateral activities were resumed under National Academy auspices. The agenda of these discussions principally addressed nuclear arms control, but also dealt with other security problems. The discussions were carried out in a collegial spirit and it was gratifying to observe that the Soviet contingent was able to express personal views more freely as time progressed. It was specifically agreed that these dialogues were not to be negotiations and would not result in written agreements or public pronouncements. Rather, the group would engage in discussions in a problem-solving spirit and then each side would be expected to report whatever transpired to representatives of their government. Meetings were held at least annually, and covered a wide range of topics including ballistic missile defense, space weapons, civil defense, and other current issues.

An essential element of the Russia dialogues was the preparation of detailed accounts covering the discussions; such accounts were widely distributed among agencies of the government. Reports were generated very rapidly by Lynn Rusten, the then CISAC staff director, who managed to take notes during the meetings adequate to provide accurate data for the reports.

In hindsight, it is difficult to assess the consequences of these discussions; they may or may not have had significant impact on ending the Cold War. However, the discussions did inject numerous ideas that then were reflected in subsequent government positions on arms control.

The CISAC bilateral discussions with the Soviets produced good personal relations or even friendships with some of the Soviet representatives. I recall a meeting in Samarkand in the 1980s when the Soviet host suggested that I address a group of students who were interested in seeing and contacting the U.S. visitors. I gave a talk describing our work and in the question and answer period following, one student asked me for whom I was going to vote in the upcoming U.S. election. I answered that in a democracy this was not a proper matter to discuss in an open meeting, and my reply was greeted by a standing ovation. Such occurrences were infrequent; most discussions remained private.

Although CISAC was chartered explicitly to support the U.S.–U.S.S.R. Interacademy dialogues, we decided to broaden the activities in two directions: to add discussions with other nations, and to conduct independent studies both to keep the membership currently involved (even though most members were doing so at any rate) and also to generate important advisory documents.

At the initiative of Frank Press, president of the Academy, a contact was established with the Chinese Academy of Sciences. That academy replied that they were enjoined from engaging in security activities but referred the contact to COSTIND, the Committee for Science and Technology for the National Defense. In turn, COSTIND organized a group of active and retired scientists to become CISAC's interlocutors, primarily from the nuclear weapons and missile establishments. That group has now become the Scientist's Group for Peace and Disarmament; notwithstanding the change in official title of the group, their membership has remained largely the same.

Meetings with this Chinese group have also been very productive, addressing arms control issues of particular interest to China.

Our discussions probably deserve credit for the fact that we provided detailed information to the Chinese on both the American and Soviet programs for so-called peaceful uses of nuclear explosions (PNEs). China had some real enthusiasts for major projects in that respect including diversion of rivers to the arid northeast of China. As a result, China had insisted on a PNE exception during the nuclear test ban negotiations in Geneva, an exception that was not acceptable to the other participants because it provides a clear loophole for military testing. China withdrew its insistence on that exception, probably prompted by the fact that the recital of the Soviet and American PNE activities in our joint meetings had documented the lack of practical utility of PNEs.

David Hamburg, then a member of CISAC, raised the question as to why we were only talking about security problems with our adversaries and not with our friends. As a result, a meeting of representatives of European academies was convened in Washington in January 1986 to discuss the potential value of expanding our dialogues to Europe. In March 1987, we organized a trip by Paul Doty, Lynn Rusten, and me to the academies of Italy, France, and the United Kingdom. I gave talks at the Accademia Nazionale dei Lincei in Rome, the Académie Française in Paris as well as at the Royal Society in London, describing the work of CISAC and its utility. We also talked in Germany, although there is no single academy in that country. Reception was generally favorable; in particular, Eduardo Amaldi in Italy "caught fire" about such broadened discussions and took the initiative to organize multinational conferences on arms control in Europe. Sadly, Amaldi died very shortly after initiating these conferences and leadership was continued by Giorgio Salvini to sponsor what became known as the Amaldi Conferences.

The utility of these conferences, which proceeded annually, was somewhat variable principally because the attendees were selected by each of the member academies, frequently for their prominence rather than their interests or experience in arms control matters; one of the explicit purposes of the meetings was to increase the competence of representatives of national academies in security matters, inasmuch as members of academies are frequently consulted by their governments on security issues. The Amaldi Conferences have been continued until very recently, but currently their future is in some doubt.

Several major studies were completed by CISAC while I was chairman. Two[1,2] dealt with U.S. nuclear weapons policy. Possibly the most important result of these studies, to me, was to clarify my thoughts on the question, "What are nuclear weapons for?" The study addressed the various forms of deterrence policy. After considerable debate, the committee agreed that the only justifiable remaining mission of nuclear weapons after the end of the Cold War would be to deter the use of nuclear weapons by others. This conclusion is equivalent to a "no first-use" policy; that is, we recommended that the United States would never be the first one to actually launch a nuclear weapon. This conclusion extends to their use to deter attacks by chemical or biological weapons. Usually nuclear, chemical, and biological weapons are lumped together under the single designation of weapons of mass destruction (WMD). I objected[3] to the use of the WMD terminology because of the fact that the military characteristics of chemical, biological, and nuclear weapons are highly dissimilar; in fact, chemical weapons do not pack any more lethal destructive power per unit of weight of delivered munitions than do "conventional" chemical explosives. Although the committee did not agree on a simple call for elimination of nuclear weapons, or even prohibition of nuclear weapons (elimination and prohibition are not the same!) , the committee pushed for the creation of conditions that would make prohibition of nuclear weapons possible. I discuss my views on these matters further at the end of this chapter.

An important contribution was the committee's reports on management and disposition of excess weapons' plutonium. It has been widely recognized that the stockpiles of plutonium accumulated in the world vastly exceed any reasonable need, and constitute a burden in terms of the risk of nuclear weapons proliferation or radioactive contamination. Moreover, production of plutonium continues as a result of reprocessing of spent reactor fuel. Prior to our report, the Department of Energy had proposed construction of reactors specially designed for the plutonium disposition mission. The committee rejected that expensive and slow option and examined all possible disposition options, ranging from such extremes as shooting the plutonium into the sun on a rocket to burying the plutonium in very deep, multikilometer long shafts. Two disposition options were recommended. The first was fabricating the plutonium into mixed oxide fuel (MOX), a combination of uranium and plutonium oxides, and then burning the MOX in existing reactors in the United States and Russia. The second option was to mix the excess plutonium with highly radioactive fission products made available from spent nuclear fuel and immobilizing this mixture by vitrifying it in heavy glass modules. This product would become essentially impervious to theft due to its radioactivity, and could be disposed of together with other forms of spent fuel in a geological repository.

In agreeing on this report, the Committee created and recommended the application of two concepts: the "spent fuel standard" and the "nuclear weapons standard." The spent fuel standard implied that the plutonium-containing material, once disposed of, should not be any more amenable to theft or diversion than ordinary spent fuel from light water reactors. The nuclear weapons standard implied that plutonium, whether separated for civilian or military purposes, should be as securely protected before it is irreversibly combined with radioactive materials as is an assembled nuclear weapon. A separate study by CISAC examined the extent to which current practices conformed to these standards.

Because the technological problems associated with the second option, that is, disposition of plutonium in a reactor, are complex, the complete study of the reactor-related options for plutonium disposition threatened to delay the issuance of the report. Accordingly, we decided to issue our report[4] and prepare a separate report, under the chairmanship of John Holdren, dealing with the reactor-related options; that separate report was issued about a year later.[5]

Both reports had a profound impact on policy but, sadly, not on eventual action. The Department of Energy reversed policy and fully adopted the recommendations. I recall personally briefing Secretary of Energy Hazel O'Leary during a one-hour automobile ride when she was returning from a visit to SLAC. Negotiations were initiated with the Soviet Union to proceed with disposition of excess plutonium, and I joined a special joint committee addressing that subject established by the academies of the two countries. Unfortunately, practical implementation of these decisions proved to be slow. The Russians were not interested in the "immobilization option" of burying plutonium as waste combined with radioactive material, but agreed to pursue the MOX disposition path. The Soviets objected to the "immobilization option" as "throwing away" the fruits of hundreds of thousands of hours of socialist labor dedicated to producing the plutonium. The Department of Energy endorsed the dual approach and initiated programs towards both methods of disposition. Presidents George W. Bush and Vladmir Putin signed a Plutonium Management and Disposition Agreement (PMDA) in the year 2000 to dispose of 34 tons of excess material, enough for about 10,000 nuclear weapons, but only about one-fifth of the total Russian inventory.

At the time of this writing, the amount of actually disposed plutonium has been zero; notwithstanding the Presidential agreement and provision of funds, the process was sidetracked by disagreements about such ancillary issues as liability of participating contractors and of access to the disposition process by personnel of the contracting parties. Costs rose dramatically, and most recently, the Russians proclaimed a loss of interest in using MOX in existing light water reactors (of which they have seven, and Ukraine has several more); instead they have proposed burning the plutonium in the one fast-neutron reactor existing in Russia, with the hope of converting our desire for plutonium disposition into American subsidies for the construction of additional fast-neutron reactors. All these vacillations have thus forestalled any progress, and the vast plutonium stockpiles continue to be in need of babysitting.

In 1983, President Reagan delivered his famous "Star Wars" speech in which he proposed an intense research and development program to make nuclear weapons "impotent and obsolete" by a defensive anti-missile shield. I had received a few days' advance notice of the speech, but there had certainly been no prior consultation with any independent members of the scientific community.

Edward Teller had been the principal advocate of Star Wars, with direct access to the President. He proposed a laser energized by a nuclear weapon which then would generate a highly directional x-ray beam as the means of providing the required shield.

Needless to say, for a number of reasons, President Reagan's speech was very confusing if not upsetting to those independent scientists who had long experience with the offense-versus-defense issue. First, the standards that a defense would have to meet to intercept essentially all incoming missiles were extremely high and, notwithstanding the conjectured laser, no realistic technology was in view, nor is in view today, to meet these standards. Second, nuclear weapons can be delivered to the U.S. homeland by many means other than ballistic missiles, so erecting a leakproof defense, or "astrodome" as it was dubbed, against only long-range ballistic missiles would have had limited value, even if it were feasible.[6] Third, as discussed previously, any defense against nuclear weapons—in particular in the U.S.–U.S.S.R. context—would be escalatory if the cost to defeat such a defense by amplified offensive measures such as decoys proved to be much cheaper than the cost of the defense. I got drawn into a large number of discussions, including congressional testimony, in which strong criticism of the Star Wars proposal was expressed.

One interesting byproduct of this role was my participation in a special study of the Pontifical Academy initiated by Pope John Paul II in late January 1985, ostensibly convened to discuss the weaponization of space, but which actually addressed the ballistic missile defense issue. A representative of President Reagan had requested that the Pope endorse the Star Wars initiative using the simplistic but defective argument that defensive measures are more conducive to peace than the offensive balance then in existence. The Pope shrewdly did not accept that request for endorsement, but instead asked his Pontifical Academy for counsel. I participated in the consequent deliberations, which advised the Pope that things weren't as simple as that, and the Pope never endorsed Reagan's proposal.

I had previously been drawn into arms control presentations on behalf of religious bodies. I gave a sermon on the nuclear arms race in New York at the Cathedral of St. John the Divine in 1982. I had some trouble preparing that talk, because I was advised that a sermon must never contain more than three ideas, a standard difficult to meet and which I thoroughly violated. I also participated in a later meeting on "Sustainable Development" convened by the Pontifical Academy, in which I discussed the economic impact of military arms races. In this context, I emphasized that although most economic activities designed to satisfy a given demand would eventually saturate that demand, in contrast military activities stimulate an increase in demand for such activities in response to the reaction of potential opponents.

Beyond participating in discussions, lectures, or testimony on ballistic missile defense and the nuclear arms race, I have also had the opportunity to engage in other aspects of national security. In 1982, the National Academy of Sciences organized a major study on scientific communication and national security chaired by Dale Corson, the president of Cornell University. At that time, charges were made that free scientific communications across national boundaries were leading to a "hemorrhage" of valuable national secrets, and therefore restraints on such communications should be enacted. Such charges were leveled in particular by the deputy director of the CIA. The NAS committee held several meetings; notwithstanding its highly varied composition, the committee reached some firmly agreed-upon conclusions.[7] I characterized any restraints on scientific communication to be analogous to hunting for leaks in a vacuum system with the valves open, because international commerce resulted in enormously more technology transfer than did scientific communication.

The principal recommendation by the panel was that the only restraint on scientific communication should be formal security classification, which had a well-established meaning and tools for implementation. Interestingly enough, the recommendation of the Corson Panel was fully accepted by the administration and resulted in the issuance by President Reagan of NSDM 189, a National Security Decision Memorandum, which specifically endorsed the recommendation that security classification should be the only restraint on scientific communication; various gray-area restraints would be counterproductive to the free exchange of ideas. That decision memorandum has been reaffirmed by all succeeding administrations but, unfortunately, efforts designed to stifle free exchange of purely scientific communication continue to resurface, and are greatly expanding.

A nuclear weapons risk of continuing concern is nuclear proliferation. An unprecedented success in arms control was the signature of the nuclear nonproliferation treaty that came into force in 1970. Over time it has been signed and ratified by all nations on the globe excepting Israel, India, and Pakistan; North Korea signed, but later withdrew from the treaty. Stemming proliferation of nuclear weapons appears essential to avoid an uncontrollable distribution of nuclear weapons, which would indeed endanger the future of civilization. Nevertheless, this task is an exceedingly difficult one: proliferation of any new technology, once invented, has never been prevented in the history of humanity.

Israel acquired nuclear weapons during my tenure on the President's Science Advisory Committee. The increasing evidence was incontrovertible, and I was troubled by the fact that on the political level the decision was apparently reached not to react to this instance of nuclear proliferation.

On September 22, 1979, a light flash was observed by one of the Vela satellites that had been orbiting for about a decade as part of the satellite family deployed specifically to monitor nuclear explosions in the atmosphere and in space. The detected flash occurred over the South Atlantic or the Indian Ocean, and the U.S. government's intelligence services were apparently persuaded that it originated from a surface nuclear explosion. As a result, President Carter convened a special eight-member panel chaired by the experienced MIT physicist Jack Ruina; I was a member of this panel.

We reviewed the available evidence with representatives of the relevant government agencies and issued a report in the spring of 1980. In my view, the evidence for the light flash originating from a nuclear weapon was unpersuasive or "not proven" to use the Scotch Verdict terminology; the report made the perhaps even stronger statement that the light flash was probably not caused by a nuclear explosion.

The Vela satellite carried two photoelectric detectors, each of which recorded a double-humped growth of light intensity as a function of time. That type of signal indeed corresponds to what is expected from a nuclear explosion: the first bump originates from the intense ionization produced by the nuclear radiation; that light signal is then dampened by ion recombination induced by the subsequent shockwave, and then reappears as the recombination dissipates. However, there were troubles presented by interpreting the data in this way.

The intensity and also the timing of the signals from the two photo tubes did not match the readings expected from a surface-based nuclear explosion, and one had to postulate a specific malfunction to make the two signals consistent with such an event. More important, examination of the records of previous signals showed that there was not a sharp break between the suspect signal and preceding smaller recordings of unknown source, which Luis Alvarez dubbed "Zoo Events," a term customarily applied to bubble chamber photographs recording tracks of unknown origin. The question naturally arose: if it is not a nuclear event, what is it? We did not answer that question, nor did it necessarily require an answer. However, speculation showed that reflected sunlight from a meteorite that might have struck the satellite could have caused a similar type light curve, but the probability of such an event appeared miniscule.

An attempt was made to corroborate the Vela light signals by other observations. The Naval Research Laboratory conducted a survey of signals seen by underwater sound detectors and indeed saw some disturbances at the day in question, but unfortunately no controlled searches were initiated to examine whether similar anomalous signals might also not have been present at other dates.

As amusing incident originated from a DOE-funded research installation that examined sheep thyroids harvested from New Zealand. It was reported that indeed some of these thyroids showed an elevated level of radioactivity. Sheep are excellent vehicles for concentrating a contamination by radioactive iodine, because they graze over large areas of potential exposure, and the iodine a sheep ingests concentrates in its thyroid. However, when visiting the research installation in question, we found the detector used to analyze the sheep thyroids to be completely unshielded, and it was further reported that elevations in counting rates from that detector were not only due to contaminated specimens, but would also be triggered by the packages of passers-by! Unfortunately, collection by aircraft of potential airborne radioactive debris from the suspected explosion was delayed by several days, and provided no evidence; it had rained heavily in the interim, so that such debris, if it had existed, could have been "rained out."

Another piece of supposedly corroborating evidence originated from the radioastronomical observatory at Arecibo, Puerto Rico. An ionospheric disturbance was reported at the time of the explosion, but a detailed analysis indicated that the direction of that disturbance was inconsistent with an origin from the suspect location.

The evidence from the 1979 light flash continues to be controversial to this day. The National Security Archives records no less than 15 documents reporting on reviews of the incident. In addition, I participated in an in-depth JASON review, which again reached an inconclusive result. I was particularly annoyed by an account by Seymour Hersh[8] which claimed that the presidential panel chaired by Jack Ruina was "instructed" to issue a negative report. The membership of that panel certainly contained many individuals, such as Luis Alvarez, who are clearly "uninstructable," and I would include myself in that category. Speculations about this event contend that if it were a nuclear explosion, it might have originated from some type of collaboration between Israel and South Africa, each of which at that time had an existing nuclear weapons program. Note that South Africa has dismantled its nuclear weapons program, including five completely assembled weapons.

The totality of these experiences has persuaded me of the crucial importance of independent science advice rendered to the highest level of government. Such advice can be divided into two categories: science in government and government in science; science in government concerns scientific input to governmental policy, whereas government in science describes the financial and administrative support of science by governmental entities. I addressed both of these topics in my earlier account of the days of the President's Science Advisory Committee. Although science in government has been deteriorating ever since that period in time, government in science is still vigorous, as some of the episodes I have cited document.

Independent science advice to the government is essential simply to prevent "bad science," which occasionally is supported by government over a protracted period of time. Such advice provides an independent voice able to both "blow the whistle" on bad science and to reach receptive ears.

Examples along these lines, in some of which I participated, are abundant. A nuclear-propelled aircraft was supported by government at large expense over a protracted period of time until independent advice that such a device was simply impractical prevailed: the combination of the weight of the required reactor and that of the shielding for the pilot was an insurmountable barrier. A more recent example is the governmental sponsorship of the potential production of an isomer of hafnium as a possible nuclear explosive.

Indeed, such an isomer in a high state of angular momentum exists; should it return to its ground state, it would release more than 2 MEV of energy per atom. The government's interest in that isotope was stimulated by a series of probably wrong experimental results from a Texas experimenter who observed a signal attributed to de-exciting that isomer when irradiated by x-rays from a conventional medical x-ray tube; critical examination of the data indicates that the statistical significance of the experiment was highly marginal. Moreover, the arithmetic describing the experiment indicates that even if the results were correct, more energy would be required to de-excite the isomer than would be released. Finally, theoretical analysis indicates that although de-excitation is possible in principle, much higher incident fluxes and higher energy x-rays would be required to achieve it than were used in the actual experiment.

Subsequent experiments at synchrotron radiation laboratories failed to confirm the existence of de-excitation. The JASON group issued a negative report on the subject. Notwithstanding this contrary evidence and the fact that a chain reaction is clearly not possible, the Defense Advanced Research Project Agency (DARPA) sponsored a panel to investigate the means and costs for large-scale production of the isomer in question and lurid viewgraphs of the potential utility of a hafnium weapon were presented in support of its production. It took heroic efforts by one of my SLAC colleagues, William Herrmannsfeldt, in addition to the JASON report, to persuade Congress to kill the activity after significant funds had been spent. Had more competent independent science advice been able to penetrate to high levels, the project would not have gotten as far as it did.

In addition to PSAC, I served on a number of independent advisory bodies, including the advisory committee to the National Nuclear Security Administration (NNSA), the defense branch of the Department of Energy. But recently, this and most other of such independent advisory bodies have either been allowed to expire at the end of their designated terms, or have been abolished outright. Examples are the Secretary of Energy Advisory Board, the advisory committee to the Department of Homeland Security, and the advisory committee on arms control to the Department of State.[9] Moreover, the scientific advisory bodies still in existence, such as the President's Council on Science and Technology (PCAST) and the Defense Science Board, are having a progressively increasing fraction of their members drawn from the industrial-scientific community and therefore can be considered to be less independent.

The issue of science advice to the highest levels of governments comes to the very heart of the relation between science and government. Essentially all the questions discussed on these pages involve not only science, but also morality and politics. Much has been written about the social responsibility of the scientist. Indeed, scientists engaged in scientific work should become aware of the social consequence of their efforts. But this is easier said than done. When engaged in scientific or technical work, most performers are consumed by the substance of their work and awareness of the consequences can come much later. When a scientist becomes aware of such consequences and enters the public arena as a citizen, then he can be accused by his peers or the public of abusing what reputation he may have as a scientist in having his opinion given undue weight. Yet if he remains silent, he can be accused of being callous and not baring the consequences of his work to the public.

I have chosen the middle road between these two approaches in pursuing basic physics and creating the tools to make progress in that field possible, while at the same time dedicating a substantial effort to examining and disseminating the impact of science on human affairs, in particular in the national security area. Let me conclude by expressing my views on the current dilemma brought into the world by of nuclear weapons created through the efforts of physicists.

It is well known that nuclear weapons have multiplied the explosive power that can be carried by munitions of a given size and weight by a factor which can exceed one million. Two weapons of average explosive yield 1/20th of those in today's stockpiles killed one-quarter of a million people in Hiroshima and Nagasaki. Notwithstanding these facts, the total inventory of nuclear weapons during the Cold War grew to about 70,000, and today, even after the end of the Cold War, the number remains at somewhat below 30,000. Clearly these numbers are vastly in excess of any reasonable security needs today.

Yet these stockpiles remain, and constitute, in themselves, a threat to humanity. In consequence, after the end of the Cold War, nuclear weapons risks continue in several categories.

- The risk that some substantial fraction of the nuclear weapons stockpiles of the United States and Russia may be launched as a result of warning yielding false alarms, or through errors in communications in command and control.
- Risks that nuclear weapons might be used in a regional conflict between nuclear-armed adversaries such as India and Pakistan.
- The risk that nuclear weapons might spread to many other countries, resulting in a situation unmanageable by human institutions.
- The risk that nuclear weapons or nuclear weapons' useable materials might reach subnational terrorists and be exploded in populated areas.

I do not give a critical discussion here of these dangers beyond regretting that the physical realities of nuclear weapons have been largely submerged under the symbolism of power which they seem to represent to political leaders. Let me, instead, make some relevant historical remarks.

After World War II, the control of U.S. nuclear weapons remained largely in the hands of the Air Force, and the Air Force leadership intended to use more and larger nuclear weapons as punitive tools to threaten and, if necessary, execute massive anti-population attacks. Concurrently, a series of major studies by scientists—in which I did not participate—considered the policies that should govern nuclear weapons. One of these studies, called Project Vista, was conducted at Caltech with J. Robert Oppenheimer being a leading participant. That study advocated design and construction of smaller nuclear weapons to be used in limited tactical attacks. The Air Force response was irate, and the report was suppressed, receiving scant attention even from those few who had access to it.

Paradoxically, during the evolution of the Cold War, the situation reversed. "Deterrence" became a common description for most policies, however, that term received a wide range of interpretations. Deterrence implies that assets of a potential opponent should be held at risk to a degree sufficient to persuade him that initiation of hostilities would result in retaliation, which in turn would deny him his initial objectives and would lead to an unacceptable loss. But there remain many unanswered questions.

What is to be held at risk? How do you understand the mindset of an opponent in judging his scale of values? What is to be done if deterrence fails? Inasmuch as none of these questions have clear answers, during the Cold War successive political leaders both in the United States and the Soviet Union progressively diversified the missions that nuclear weapons were supposed to accomplish and the insane buildup mentioned above was the result.

Ironically, both scientists and members of the defense establishment reversed their initial position on the mission of nuclear weapons after World War II. Some military leaders and civilian analysts maintained that "nuclear war fighting" at a variety of levels of nuclear violence would be possible, and that if war broke out, the West should prevail in a protracted nuclear exchange or that, in other words, "a nuclear war could be won."

In contrast, most scientists now conclude that such a course would lead to uncontrollable escalation with devastating results, and that only "finite" or even "minimum" deterrence, without anticipating actual military use of nuclear weapons, would be a prudent approach. Specifically, the Cold War policy of "extended deterrence," that is, not only deterring a nuclear attack by others but also employing nuclear weapons to deter a variety of nonnuclear aggressive moves, has been largely rejected by the scientific community.

However, during the Cold War in Europe, American and NATO policies largely promoted the role of nuclear weapons in compensating for the perceived inferiority of conventional weapons in NATO forces, thereby using nuclear weapons to deter conventional aggression by the Soviet Union. Ironically, Russia today appears to justify its nuclear arsenal—the world's largest—as needed to compensate for the perceived inferiority of Russian conventional forces.

Nuclear weapons have not been used in combat for over 60 years, but will that nonuse taboo hold? The nuclear weapons risks listed above remain very real, but at the same time, nuclear weapons cannot be uninvented. Pleas for "eliminating" nuclear weapons lack reality, but "prohibiting" them is within the range of feasibility. `Prohibition" and "elimination" are not the same; note that prohibition implies the possibility of limited clandestine evasion.

Prior to possible prohibition, it seems feasible to me to drive for consensus that the only justifiable remaining role of nuclear weapons is deterrence of the use of nuclear weapons by others. Retaining, or even searching for, other missions for nuclear weapons is shortsighted and prolongs or even exacerbates the nuclear dangers. Such a restriction on the mission of nuclear weapons is equivalent to a universal declaration of "No First-Use" of nuclear weapons, a declaration which at this time has been embraced only by China, but by none of the other nuclear weapons states. But most important, such a restricted view of the mission of nuclear weapons should enable drastic reductions of the existing nuclear weapons stockpiles, in particular those held by the United States and Russia.

Such a limit imposed on the role of nuclear weapons can be used to revitalize the nuclear weapons arms-control drive, which lately has suffered a series of setbacks. This has been a personal disappointment to me because some of the achievements in nuclear weapons arms control that were enacted during the Cold War, including some which I participated in developing, have now fallen on hard times.

The U.S. government has opposed formal arms control treaties largely on the grounds that they "limit U.S. flexibility." Indeed, they do: treaties are binding agreements that transcend any one administration and preempt U.S. law. But such formal strictures are a necessity if they are to endure and become binding on all parties to the treaty. Currently, arms control is largely being promoted through enforcement by "a coalition of the willing," meaning that different standards are being applied based on whether the relevant nation is "good" or "evil." But which nation is presumed to harbor good guys or bad guys changes in time, and even depends on which nation is perceived to support or not support current U.S. interests: nuclear arms control, to be effective, must be more enduring than that. Selective enforcement, as practiced by the current administration, cannot be lasting control.

The United States, as the unquestioned leader—measured by nonnuclear armaments and economic strength—should have the strongest possible interest in leading the reining-in of nuclear weapons on an irreversible basis.

But the excessive emphasis on the part of the United States on solving international conflicts by military force is a driver toward nuclear proliferation: less prosperous nations who cannot match the conventional military prowess of the United States are driven to compensate by acquiring nuclear weapons.

Restricting the mission of nuclear weapons, followed by reducing nuclear weapons inventories by treaty and stringent control of weapons-useable material, does not in itself prohibit nuclear weapons, but still greatly reduces the above-listed dangers of nuclear weapons. Anything less than such a drastic move simply postpones the need for action. Must we wait for a nuclear catastrophe before such actions are taken?

Endnotes

Preface

[1] W. C. Sellar and R. J. Yeatman, *1066 and All That*, (Methuen, London, 1930).

Chapter 1

[1] *Erwin Panofsky Korrespondenz, herausgegeben von Dieter Wuttke*, Volume 1, 2001; Volume 2, 2003; Volume 3, 2006 (Harrasowitz Verlag, Wiesbaden).

[2] *Fast wie mein Eigenes Vaterland, Briefe 1886–1889*. Edited by Shiro Ishii, Ernst Lokowandt, and Yukichi Sakai, (Deutsche Gesellschaft für Natur-und Völlkerkunde Ostasien, Judicium Publishers, 1995).

[3] B. Donath, *Physikalisches Spielbuch für die Jugend* (Friedrich Vieweg und Sohn, Braunschweig, 1902).

[4] A "Seminar" was an established institution, a regular gathering of intellectuals.

Chapter 2

[1] Wilhelm Busch, *Die schönsten Bildergeschischten* (Gondrom Verlag, Bayreuth, 1993). Poem entitled *Plisch und Plum*, "Zugereist in diesen Gegend, noch viel mehr als sehr vermögend, in der Hand das Perspectiv, kam ein Mister namens Pief." Translation: "Having traveled into this region, even much more than very affluent, in his hand a looking glass, came a Mister by the name of Pief." Subsequently, Pief fell into a lake and was pulled out by two small dogs, Plisch and Plum.

[2] W. K. H. Panofsky, The dependence of emission on the relative amounts of barium and strontium oxides in cathode coatings, RCA Engineering Report LR-71 (September 21, 1937).

[3] W. K. H. Panofsky, The vibrations of a piano string, junior thesis, Princeton University, 1937.

[4] W. K. H. Panofsky, The construction of a high pressure ionization chamber, senior thesis, Princeton University, 1938.

Chapter 3

[1] W. K. H. Panofsky, *Jesse W. M. DuMond, Biographical Memoirs*, (National Academy of Sciences, National Academy Press, Washington, DC, 1980), Vol. 52, pp. 160–201.

[2] J. W. DuMond, W. K. H. Panofsky, and A. E. S. Green, A precision determination of h/e by means of the short wave-length limit of the continuous x-ray spectrum at 20-KV, *Phys. Rev.*, **62** (5), 214 (September 1 and 15, 1942).

[3] J. W. DuMond, W. K. H. Panofsky, and A. E. S. Green, *Phys. Rev.* **62** (1942), Fig. 4.

[4] W. K. H. Panofsky, J. W. M. DuMond, R. R. Yost, and H. A. Kirkpatrick, Design, assembly and preliminary tests of a curved crystal gamma-ray spectrometer, *Phys. Rev.* **59**, 219 (Jan. 15, 1941).

[5] J. W. M. DuMond, E. R. Cohen, W. K. H. Panofsky, and E. Deeds, A determination of the wave forms and laws of propagation and dissipation of ballistic shock waves, *J. Accoust. Soc. Am.* **18**, 97–118 (July 1946). Sept. 27, 1945.

[6] R. A. Millikan, D. Roller, E. C. Watson, C. D. Anderson, and W. K. H. Panofsky, *Electricity and Magnetism* (California Institute of Technology, 1941).

[7] For a recent scholarly account, see Kai Bird and Martin J. Sherman, *American Prometheus: the Triumph and Tragedy of J. Robert Oppenheimer* (Alfred A. Knopf, New York, 2005).

[8] Luis W. Alvarez, *Alvarez: Adventures of a Physicist*, (Basic, New York, 1987).

[9] This account given here is a brief and incomplete summary of the factors at play. For a more complete account the reader is referred to the extensive literature.

[10] See, e.g., B. Bernstein, The atomic bombings reconsidered, *Foreign Affairs*, **74** (1), 135ff (1995).

[11] For example, the extensive analysis of the competing potential factors in B. Bernstein, Roosevelt, Truman, and the atomic bomb, 1941–1945: A reinterpretation, *Polit. Sci.Quart.*, **90** (1), 23ff (Spring 1975).

[12] E. J. Lofgren, The principle of phase stability and the accelerator program at Berkeley, 1945–1954. In *Proceedings of the 50th Anniversary of the Discovery of Phase Stability Principle, Joint Institute for Nuclear Research* (Dubna P.N. Lebedev Physics Institute, Moscow, July 12–15, 1994).

Chapter 4

[1] He later became Assistant Secretary for Defense Programs in the Department of Energy.

[2] Luis W. Alvarez, Hugh Bradner, Jack V. Franck, Hayden Gordon, J. Donald Gow, Lauriston C. Marshall, Frank Oppenheimer, Wolfgang K. H. Panofsky, Chaim Richman, and John R. Woodyard, Berkeley proton linear accelerator, Radiation Laboratory, Department of Physics, University of California, Berkeley, California, *Rev. Sci. Instrum.* **26**, 111 (1955).

[3] Pierre M. Lapostolle and Albert L. Septier (Eds.), *Linear Accelerators* (North Holland, Amsterdam, 1970).

[4] L. Alvarez et al. *Rev. Sci. Instrum* **26,** 113, Fig. 2 (1955).

[5] W. K. H. Panofsky, Linear accelerator beam dynamics, University of California Radiation Laboratory Report No. UCRL-1216 (February, 1951).

[6] W. K. H. Panofsky, C. Richman, and F. Oppenheimer, Control of the field distribution in the linear accelerator cavity, *Phys. Rev.* **73**, 535(A) (1948).

[7] E. M. McMillan, *Phys. Rev.* **80**, 493 (1950).

[8] L. Alvarez et al. *Rev. Sci. Instrum.* **26**, 114, Fig. 3 (1955).

[9] W. K. H. Panofsky and M. Phillips, *Classical Electricity and Magnetism*, 2nd ed. (Addison Wesley, Reading, MA, 1955. Republished by Dover, New York, 2005).

[10] J. Lebowitz, W. K. H. Panofsky, and S. Rice, Melba Newell Phillips: Obituary, *Phys. Today* **58** (7, July), 80–81 (2005).

[11] W. K. H. Panofsky and F. Fillmore, The scattering of protons by protons near 30-MeV: 1 photographic method, *Phys. Rev.* **79**, 57 (1950).

[12] B. Cork, L. Johnston, and C. Richman, *Phys. Rev.* **79**, 71 (1950).

[13] W. K. H. Panofsky and R. Phillips, Evidence of a p,d reaction in carbon, *Phys. Rev.* **74**, 1732 (1948).

[14] W. K. H. Panofsky and F. Fillmore, *Phys. Rev.* **79**, 69, Fig. 17 (1950).

[15] R. F. Mozley, *Phys. Rev.* **80**, 493 (1950).

[16] E. Gardner and C. M. G. Lattes, *Phys. Rev.* **74**, 1236 (1948).

[17] I asked Lawrence casually how much these shielding blocks had cost. After hearing the answer, I replied that it would be cheaper to shield the cyclotron with graduate students. At the time, the graduate student assistants were in near-revolt over low pay.

[18] R. Bjorklund, W. E. Crandall, and H. F. York, *Phys. Rev.* **77**, 213 (1950).

[19] W. K. H. Panofsky, R. L. Aamodt, and J. Hadley, *Phys. Rev.* **81** 566, Fig. 1 (1951).

[20] W. K. H. Panofsky, R. L. Aamodt, J. Hadley, and R. Phillips, The gamma-ray spectrum resulting from capture of negative pi mesons in deuterium, *Phys. Rev.* **80**, 94 (1950).

[21] W. K. H. Panofsky, R. L. Aamodt, and J. Hadley, The gamma-ray spectrum resulting from capture of negative pi mesons in hydrogen and deuterium, *Phys. Rev.* **81**, 565 (1951).

[22] W. K. H. Panofsky, R. L. Aamodt, and J. Hadley, *Phys. Rev.* **81,** 566, Fig. 2 (1951).

[23] W. K. H. Panofsky, R. L. Aamodt, and J. Hadley, *Phys. Rev.* **81,** 571, Fig. 12 (1951).

[24] W. K. H. Panofsky, R. L. Aamodt, and J. Hadley, *Phys. Rev.* **81,** 573, Fig. 15 (1951).

[25] A. S. Wightman, *Phys. Rev.* **77**, 521 (1950).

[26] W. Blocker, R. Kenney and W. K. H. Panofsky, Transition curves of 330-MeV Bremsstrahlung, *Phys. Rev.* **79**, 419 (1950).

[27] W. K. H. Panofsky, J. Steinberger, and J. Steller, Evidence for the production of neutral mesons by photons, Phys. Rev. **78**, 802 (1950).

[28] W. K. H. Panofsky, J. Steinberger, and J. Steller, *Phys. Rev.* **78,** 804, Fig. 3 (1950).

[29] See, e.g., J. Steinberger and A. S. Bishop, *Phys. Rev.* **78**, 494 (1950).

[30] J. Steinberger, *Learning About Particles – 50 Insightful Years*, (Springer, New York, 2004).

[31] E. A. Martinelli and W. K. H. Panofsky, The lifetime of the positive pi meson, *Phys. Rev.* **77**, 465 (1950).

[32] W. K. H. Panofsky and W. R. Baker, A focusing device for the external 350-Mev proton beam of the 184-inch cyclotron at Berkeley, *Rev. Sci. Instrum.* **21**, 445 (1950).

[33] K. Crowe, W. K. H. Panofsky, R. H. Phillips, and D. Walker, Precision analysis of gamma ray spectra from high-energy proton collisions with nuclei, *Phys. Rev.* **83**, 893 (1951).

Chapter 5

[1] There exists a large literature on all the events mentioned. For the most recent and very comprehensive coverage, see Chapter 3, Footnote 7. For the most detailed coverage giving a detailed sequential account, see Herbert F. York, *The Advisor*, (Stanford University Press, Stanford, CA, 1976; reissued 1984).

[2] One pure fission (nonthermonuclear) test explosion in 1952 had a yield near 500 kT (!) compared to 13 kT of the Hiroshima bomb.

[3] L. H. Thomas, *Phys. Rev.* **54**, 580 (1938).

[4] Particles entering the fringe field of a solenoid off-axis acquire a transverse component of their motion; that transverse component in turn interacts with the longitudinal field in the solenoid to produce a force towards the axis. Thus the focusing strength of a magnetic solenoid is a "second order" effect and varies as the square of the magnetic field intensity.

[5] I commemorated Lawrence's decision by a poem, with apologies to E. A. Poe.

The Meeting

W. K. H. Panofsky with help from Polly Gow, 1950.

Once upon a morning dreary,
While I pondered weak and weary
Over many a quaint and curious bit of Physics lore–
Even so, my heart was singing,
"Til there came a sudden ringing.
What message could that phone be bringing?
A signature just for the store?
Or could it be something more?

I aroused myself on answering bent,
"Hello," I said, into the instrument.
A mellow voice spoke with intent,
"Your forgiveness I implore–
But the fact is there's a meeting,
And your presence I'm entreating,
At eleven, says the Chieftain*
To discuss one problem more."
Ah yes, meetings evermore.

At the hour thus appointed,
not to have Him disappointed,
Hurried I then double-jointed,
swiftly to the Chieftain's door.
The Chief in the center, the others flanked 'im,
All he summoned, and he thanked 'em,
As I tiptoed across the sanctum
Noiseless on the tufted floor
Late – as every time before.

The controversy now was burning;
for the reason I was yearning.
Now – the purpose I was learning.
Horror struck me to the core!
Yes, my ears had heard it rightly,
(How could He regard it lightly?)
But He nodded, smiling brightly,
Quoth the Chieftain, "LIVERMORE."
Only this and nothing more.

*E. O. Lawrence

Chapter 6

[1] C. Stewart Gillmor, *Fred Terman at Stanford: Building a Discipline, a University, and Silicon Valley* (Stanford University Press, CA, 2004).

[2] Edward L. Ginzton, *Times to Remember: the Life of Edward L. Ginzton* (Blackberry Creek Press, Berkeley, CA, 1995).

[3] M. Chodorow, E. L. Ginzton, W. W. Hansen, R. L. Kyhl, R. B. Neal, and W. K. H. Panofsky, Stanford high-energy linear electron accelerator (MARK III), *Rev. Sci. Instrum.* **26**, 134 (1955).

[4] Richard B. Neal, A high energy linear electron accelerator, M.L. Report No. 185 (Linear Electron Accelerator Project, Microwave Laboratory, Stanford University, CA, 1953).

[5] En Lung Chu, The theory of linear electron accelerators, Technical Report, Linear Electron Accelerator Project, ML Report No. 140 (Microwave Laboratory, Stanford University, CA, May 1951).

[6] W. K. H. Panofsky, Achromatic translation system for high-energy beams, *Phys. Rev* **93**, 949 (1954).

[7] W. K .H. Panofsky and J. A. McIntyre, Achromatic beam translation system for use with linear accelerators, *Rev. Sci. Instrum.* **25**, 287 (1954).

[8] M. Chodorow et al., *Rev. Sci. Instrum.* **26**, 196, Fig. 6.1 (1955).

Chapter 7

[1] W. K. H. Panofsky and D. Reagan, The reaction N-14 (gamma, 2n) N-12, *Phys. Rev.* **87**, 543 (1952).

[2] G. B. Yodh and W. K. H. Panofsky, Pion production by inelastic scattering of electrons in hydrogen, *Phys. Rev.* **105**, 731 (1957).

[3] A. J. Lazarus, W. K. H. Panofsky, and F. R. Tangherlini, Photoproduction of positive pions in the angular range 7-degrees<theta (center-of-mass) <27-degrees and photon energy range 220-MeV<K<390-MeV, *Phys. Rev.* **113**, 1330 (1959).

[4] R. A. Alvarez, K. L. Brown, W. K. H. Panofsky, and C. T. Rockhold, Double focusing 0 dispersion magnetic spectrometer, *Rev. Sci. Instrum.* **31**, 556 (1960).

[5] W. K. H. Panofsky and E. A. Allton, The form-factor of the photopion matrix element at resonance, *Phys. Rev.* **110**, 1155 (1958).

[6] W. K. H. Panofsky and E. A. Allton, *Phys. Rev.* **110,** 1162, Fig. 7 (1958).

[7] G. E. Masek and W. K. H. Panofsky, Evidence for the electromagnetic production of mu mesons, *Phys. Rev.* **101**, 1094 (1956).

[8] W. K. H. Panofsky and G. W. Tautfest, Measurement of the radiative correction to electron-proton scattering by observation of the absolute cross-section, *Phys. Rev.* **105**, 1356 (1957).

[9] D. M. Bernstein and W. K. H. Panofsky, Bremsstrahlung yield of high-energy electrons in hydrogen, *Phys. Rev.* **102**, 522 (1956).

[10] W. K. H. Panofsky and W. Wenzel, Some considerations concerning the transverse deflection of charged particles in radiofrequency fields, *Rev. Sci. Instrum.* **27**, 967 (1956).

[11] P. R. Phillips, Microwave separator for high energy particle beams, *Rev. Sci. Instrum.* **32**, 13 (1961).

[12] L. N. Hand and W. K. H. Panofsky, Magnetic quadrupole with rectangular aperature, *Rev. Sci. Instrum.* **30**, 927 (1959).

[13] W. K. H. Panofsky and A. Saxena, Search for enhancement of bremsstrahlung produced by 575-MeV electrons in a single crystal of silicon, *Phys. Rev. Lett.* **2**, 219 (1959).

[14] G. K. O'Neill, W. C. Barber, Burton Richter, and W. K. H. Panofsky, A Proposed Experiment on the Limits of Quantum Electrodynamics (High-Energy Physics Laboratory, W.W. Hansen Laboratories of Physics, Stanford University, CA, May 1958).

[15] K. M. Crowe, R. H. Helm, and G. W. Tautfest, Preliminary data on the measurement of the μ+ - ß+ decay spectrum," *Phys. Rev.* **99**, 872 (1955).

[16] Proceedings of the Sixth Annual Rochester Conference, April 3–7, 1956, p. IX, 47–48.

[17] W. K. H. Panofsky, V. L. Fitch, R. M. Motley, and W. G. Chestnut, Measurement of the total absorption coefficient of long-lived neutral K particles, *Phys. Rev.* **109**, 1353 (1958).

[18] A parody from Shakespeare's Othello: Iago says, "Who steals my purse steals trash . . . but he that filches from me my good name Robs me of that which not enriches him And makes me poor indeed."

Chapter 8

[1] Edward Teller expressed his personal view in his syndicated newspaper article: "Testing in space provides a loophole through which one could drive a herd of elephants." *Houston Post*, (August 16, 1960).

[2] Harold K. Jacobson and Eric Stein, *Diplomats, Scientists and Politicians: the United States and the Nuclear Test Ban Negotiations* (University of Michigan Press, Ann Arbor, 1966).

[3] James R. Killian, Jr., *Sputnik, Scientists, and Eisenhower: A Memoir of the First Special Assistant to the President for Science and Technology* (MIT Press, Cambridge, MA, 1977).

[4] George B. Kistiakowsky, *A Scientist at the White House: The Private Diary of President Eisenhower's Special Assistant for Science and Technology* (Harvard University Press, Cambridge, MA, 1976).

[5] Walter A. Rosenblith, *Jerry Wiesner, Scientist, Statesman, Humanist: Memories and Memoirs* (MIT Press, Cambridge, MA, 2003).

[6] Sub-Committee of the President's Science Advisory Committee, appointed by President John Kennedy on June 28, 1961, headed by W. K. H. Panofsky; other members of the panel were William O. Baker, Vice President, Bell Telephone

Laboratories; Hans A. Bethe, Professor of Physics, Cornell University; Norris E. Bradbury, Director, Los Alamos Scientific Laboratory; James B. Fisk, President, Bell Telephone Laboratories; John S. Foster, Director, University of California Radiation Laboratory; George B. Kistiakowsky, Professor of Chemistry, Harvard University; Frank Press, Director, Seismological Laboratory, California Institute of Technology; Louis H. Roddis, President, Pennsylvania Electric Co.; John W. Tukey, Professor of Mathematics, Princeton University; and Walter H. Zinn, Vice-President, Nuclear Division, Combustion Engineering, Inc.

[7] National Academy of Sciences, *Technical Issues Related to the Comprehensive Nuclear Test Ban Treaty* (National Academy Press, Washington, DC, 2002).

[8] I noted that during my membership in PSAC, I also led SLAC, supported by U.S. government funding.

[9] Eisenhower referred to the PSAC members as "my scientists," and deliberately preferred ignorance on their partisan preferences.

Chapter 9

[1] R. B. Neal and W. K. H. Panofsky, The Stanford MARK III linear accelerator and speculation concerning the multi-GeV applications of electron linear accelrators, Technical Report #80 (High Energy Physics Lab, Stanford University, Stanford, CA) April, 1956.

[2] Richard B. Neal, General Editor, *The Stanford Two-Mile Accelerator* (W.A. Benjamin, New York, 1968).

[3] C. Stewart Gilman, *Fred Terman at Stanford* (Stanford University Press, CA, 2004), pp. 361 ff.

[4] Stanford University proposal for a two-mile linear electron accelerator (Stanford University, Stanford, CA, April, 1957).

[5] George B. Kistiakowski, A Scientist at the White House: The Private Diary of President Eisenhower's Special Assistant for Science and Technology (Harvard University Press, Cambridge, MA, 1976).

[6] Address by the President at the Symposium of Basic Research. Sponsored by The National Academy of Sciences, The American Association for the Advancement of Science and the Alfred P. Sloan Foundation (May 15, 1959).

[7] Hearings before the Subcommittees on Research and Development and on Legislation of the JCAE (July 14 and 15, 1959).

[8] Record of JCAE Hearings, FY '61.

Chapter 10

[1] Richard B. Neal, General Editor, The Stanford Two-Mile Accelerator (W.A. Benjamin, New York, NY, 1968).

[2] K. R. Trigger, SLAC –TN-64-19 (1964).

[3] W. K. H Panofsky and M. Bander, Asymptotic theory of beam breakup in linear accelerators, *Rev. Sci. Instrum.* **39**, 206 (1968).

[4] Adele Panofsky, Stanford paleoparadoxia fossil skeleton mounting, SLAC Pub. 7829 (September 1998).

[5] Since Alvarez was now preaching like a clergyman, he should dress like one as well.

Chapter 11

[1] SLAC construction began in July, 1962 and was completed in January, 1967.
[2] S. M. Keeny, Jr., and W. K. H. Panofsky, MAD vs. NUTS, *Foreign Affairs*, **60**, 287–304, (1981).

Chapter 13

[1] D. H. Coward et al., Electron–proton elastic scattering at high momentum transfers, *Phys. Rev. Lett.* **20**, 292 (1968).
[2] L. Hoddeson, L. Brown, M. Riordan, and M. Dresden (Eds.), *The Rise of the Standard Model*, (Cambridge University Press, Cambridge, UK, 1997).
[3] W. K. H. Panofsky, Electromagentic interactions: Low q2 electrodynamics: Elastic and inelastic electron (and muon) scattering, in *Proceedings of the Fourteenth International Conference on High Energy Physics* (CERN, Geneva, 1968), pp. 23–39.

Chapter 14

[1] I do not recall this meeting, but Richter definitely confirms that it took place.
[2] L. Hoddeson et al. (Eds.), *The Rise of the Standard Model* (Cambridge University Press, Cambridge, UK, 1997), p. 62, Fig. 4. 3; (revision of J. E. Augustin et al., *Phys. Rev.* **23**, 1407, Fig. 1a (1974).
[3] W. K. H. Panofsky, Particle discoveries at SLAC, *Science, New Series*, **189**, 1045 (September 26, 1975).

Chapter 15

[1] A. I. Johnston, W. K. H. Panofsky, M. di Capua, and L. R. Franklin, *The Cox Committee Report: An Assessment*, edited by M. M. May (Center for International Security and Cooperation, Stanford University, December 1999).

Chapter 16

[1] M. Riordan, The demise of the superconducting supercollider, *Physics in Perspective*, **2**, 411 (2000).
[2] M. Riordan, A tale of two cultures: Building the superconducting supercollider, 1988–1993, *HSPS*, **32**, 125 (2001).
[3] G. F. Dugan and J. R. Stanford (Eds.), Superconducting super collider: A retrospective summary, 1989–1993. Superconducting Super Collider Laboratory Report, SSCL-SR-1235, (April 1994).
[4] I remember when the issue of the "critical size" of a collaboration was discussed at a meeting of a physics survey committee of which I was a member, convened by the

National Academy of Sciences. The great theoretical physicist Eugene Wigner, also a member, said in his customary restrained voice, "May I add a footnote: 'Sometimes contributions are being made by individuals.'"

Chapter 17

[1] The Committee on International Security and Arms Control, *The Future of Nuclear Weapons Policy* (National Academy Press, Washington, DC, 1997).

[2] The Committee on International Security and Arms Control, *The Future of the U.S. Soviet Nuclear Relationship* (National Academy Press, Washington, DC, 1991).

[3] W. K. H. Panofsky, Dismantling the concept of 'weapons of mass destruction', *Arms Control Today* **28**, 3 (1998).

[4] Committee on International Security and Arms Control, *Management and Disposition of Excessive Weapons Plutonium* (National Academy Press, Washington DC, 1994).

[5] Committee on International Security and Arms Control, *Management and Disposition of Excess Weapons Plutonium: Reactor Related Options* (National Academy Press, Washington DC, 1994).

[6] W. K. H. Panofsky and D. Wilkening, in *US Nuclear Weapons Policy*, edited by G. Bunn and Christopher F. Chyba (Brookings Institution Press, Washington, DC, September 2006).

[7] Panel on Scientific Communication and National Security, Committee on Science, Engineering, and Public Policy, The National Academies, *Scientific Communication and National Security* (National Academies Press, 1982).

[8] Seymour Hersh, *The Samson Option: Israel's Nuclear Arsenal and American Foreign Policy*, (Random House, New York, 1991), pp. 280–281.

[9] More precisely, the Arms Control and Nonproliferation Advisory Board (ACNAB). Its function lapsed for five years but recently it has been reinstated with new membership.

Subject Index

Printed in Thailand